CIVIL ENGINEERING SOLUTIONS

AN INNOVATIVE GUIDE TO ADVANCED CIVIL ENGINEERING CONCEPTS

CIVIL ENGINEERING SOLUTIONS

AN INNOVATIVE GUIDE TO ADVANCED CIVIL ENGINEERING CONCEPTS

PREM VARDHAN

Old No. 38, New No. 6
McNichols Road, Chetpet
Chennai - 600 031

First Published by Notion Press 2016

ISBN 978-1-945400-74-2

Contents

Preface

Everyone aspires to do something in order to gain recognition in society. This is a challenge. Some new challenges crop up while implementing the assigned or chosen task.

We make an effort to create an atmosphere that can help one develop their desire to do something new. This is nurturing one's innovative senses. With small successes, it grows manifold and looks forward for more opportunities with interest and confidence.

I have selected major activities as many as could be possible for construction of projects like reclamation, roads, bridges, dams, sea ports, airports, industrial plants and buildings. Innovative concepts and simple procedures are taught for execution of these activities. These methods would further be guide to resolve situations not covered in the book.

Every parent wishes that their child excels in life better than themselves. This is taken care of in a small way by counseling parents, schools and teachers.

Wishing all young engineers and other enterprising people the greatest prosperity in their careers and life.

Prem Vardhan

Chartered Civil Engineer

vardhanprem@yahoo.com

Acknowledgements

This work is respectfully dedicated to my teachers

Mr & Mrs Maharaja Krishna, my parents

Mr Shashi Ruia, Chairman, ESSAR Group

Mr. H. D. Singh, Director, Continental Constructions Ltd.,

Mr. D.N. Sood, Zonal Manager, Gammon India Limited

I thank and appreciate all the authors and photographers whose photographs and captions I have picked up from the internet for the sake of illustrating history.

LEVEL 1

Introduction to Civil Engineering

1.1 Who is an Engineer?

A supernatural power has engineered a process to create the universe and life. An engineer is one who creates something by a process called the systemization of knowledge, will, wisdom, passion and tools. As a matter of fact, engineering is common sense with a flavor of expert knowledge specific to the subject. The one who practices engineering on a daily basis by creating or making something as a routine or even better in a systematic manner is called an engineer. He may have specialization in the particular subject.

When someone writes on a piece of paper, he is able to write because he knows the process of writing. His hand and brain are trained to write the script, make words and sentences in order to express himself, while others who have not been trained in this knowledge cannot write. His teacher has engineered the training while imparting this knowledge to him. The quality of writings depends upon the extent of knowledge and expertise that are gained from training and experience.

The preparation of food we eat is a process engineered by cook or chef. Her or his passion for innovations arouses our senses of taste, sight and smell.

Everybody lives in a shelter of some kind. Houses need electricity for light, heating, cooling and running the appliances. This has reached the present stage and quality after centuries of engineering research.

We use many equipment and appliances like refrigerators, washing machines, bicycle, motorbikes, cars and such else. These things have come after very long drawn efforts of engineering research in creating concepts, utility, design, manufacture, sales, delivery and even service at the doorstep.

Engineering is studied through the following main branches:

- Civil engineering
- Electrical engineering
- Mechanical engineering
- Chemical engineering
- Metallurgical engineering

Now, there are many new branches that have come into existence.

In this book, we will concentrate on civil engineering and pick up portions of expertise from other branches as necessary. We will look into ways for making our children successful first, and then guide all those who are interested in civil engineering in their training and professional careers.

A civil engineer designs, builds and maintains houses and buildings of all types, roads and bridges, dams, railways, air and sea ports, and develops means to provide civic amenities including town-planning and their layout, with market places, community centres, water, sanitation, drainage, pathways and such else.

Let's go back in history of civil engineering and its many achievements.

1.2 Shelters - Homes for Us

There has always been a continuous effort to create something better and newer to satisfy our basic needs. Shelter has evolved through innovative thoughts and ideas. As a result, we are in the present stage of comfort, and still continue to exert ourselves to find ways to improve our lives. The words '**satisfaction**' and '**impossible**' are nonexistent in the dictionary of engineers. The day this happens, progress in the whole world comes to a halt.

Prehistoric Houses

Caves

In prehistoric times humans in cold climate used to live in caves and slowly graduated to tents made of animal thick skins and bones.

In forests, they made houses on tree tops by jungle produce with bows and arrows for hunting.

Slowly they improved their living. They saw new plants grow under the old plants due to dropping of seeds on the ground. They picked up preferred plants, collected seeds and grew new plants in fairly cleaner area, and this gave birth to agriculture and they could have some choice in their food. This happened around 6000 BC

4000 BC

Humans always had desire to improve their living conditions by 6000 BC they had some choice on food by growing plants of their choice selected from surround and the next 2000 years they started heating and roasting food before eating, developed mud mortar and started making walls of their houses with stones and mud mortar with thick grass to protect from heat and rain on supports of wooden logs for the roof.

With passage of time, with increase in living comforts, their life style also changed as it is evident now and per force their body and life style started asking for more and more and accomplished with persistent thinking, stressing their mind for innovative ideas for betterment in living.

Around 2600 BC, Indus Valley Civilization

City of Mohenjo-Daro came in existence in around 2600 BC as one of the major known civic developments popularly known as Indus Valley Civilization. Streets were well laid and houses were made of small burnt brick in two to three stories. Floors in upper stories were big stone slabs laid on dressed wooden logs, acting as beams. They had effective drainage system and reasonable good town-planning with central assembly place, streets and well laid houses. Even they had their own weights and measures. However, this must have happened after a prolonged development process.

Floor slabs of dressed stones on timber or steel beams are still seen in India in buildings about one hundred years old, still strong and everlasting.

Along with Indus Valley Civilization, development of other civilizations in world were a big land mark in development of civic activities, comforts and construction technology, there was an all-round development in the world for the next 2000 years and new concepts were evolved very rapidly. Basic amenities like drainage, streets and multistoried houses with hygienic toilets were developed.

There were many innovations in good architecture, better construction of big structures.

Remarkable developments took place all over the world, name a few, Catal Huyuk, now known as turkey, Sumer in Iraq, Egypt, Greece, Persia, France, British Isles etc. were pioneers in development.

All the nobles and kings started promoting quality and big houses for themselves and their close associates in a good race to have better living than others. Many beautiful and durable forts, palaces and houses were built all over the world. Civil engineering technology also gained momentum and developed much faster than before in this race of building something better than others, forced engineers and architects to think and develop new innovative concepts in architecture and construction.

Readers are advised to go to internet and get information and photos of such buildings and important structures built and improved progressively with passage of time.

Around 324 BC

Around 324 BC Maurya Empire was established in India built many forts, temples and monuments a big leap to civil engineering knowledge understanding and implementation.

Great wall of China construction started around 210 BC with local material just to keep people occupied and avoid them to becoming lazy.

By this time civilization around the world had developed substantially and people started living in good manmade houses, with market places and magnificent forts, palaces and places of worship. Remains of this civilization and buildings are seen all over the world. Their architecture and good construction is generally surprising and was possible since time and money were secondary consideration or rather no consideration, except desire and determination to build something unique, un parallel in the world and everlasting. This continued up to around1800 AD. All the wonderful civil engineering structures were built in this period.

On the contrary all creatures other than humans on this planet does not have such senses of innovations and desire for improvement of their living. However, making a hut in prehistoric ages by humans might be by inspirations drawn from birds making their nests for laying eggs and protection of their children.

Otherwise animals still live as they were long ego, except cases when manmade comforts are provided by humans. Interestingly pet dog, birds, cat and cattle catches cold on getting wet, while those in forests nothing happens for such change in climate conditions. Pets become used to comfortable life and small change of environment is not tolerated and they become sick.

Taj Mahal in Agra, India was built during 1632–1650 AD. Acoustic design and management for many forts, including, main assembly hall in Agra fort is still difficult to repeat. There are still existing many forts where on entry at gate by a visitor in bottom of the fort, his footsteps would be echoed to the king sitting on top of hill in the fort.

One could see a big heap of soil near old and tall structures. A ramp on one side of the structure was built and raised as building progressed for taking to top of structure, big carved stone blocks and other building material through these ramps on trollies pulled by, bullocks, horses and slaves. Ramps were dismantled after finish of work and soil deposited in a heap nearby.

Up till this time as earlier stated civil engineering is common sense with right applications and people with vision, fascination and right perceptions kept on building such structures, still considered magnificent. The knowledge and art was transferred to youngsters by teachers, however both student and teacher used to have good understanding, respects and hence bonding.

Slowly, demand was increasing for talented engineers and teachers had to part their knowledge, to cope up with requirement of society to big groups and school concept was formed, which is called GURUKUL in India.

In Europe and United States, these Guru Kul were called engineering colleges and group of colleges with teaching different talents were called Universities.

This engineering college concept spread all over the world and students started going to premier engineering colleges oversees and later spread the knowledge in their own country, even with the help of a few experts on deputation from oversees.

When Archimedes invented floatation while taking bath, ran necked in streets, shouting eureka. Similarly, someone invented light bulb, heat engines, magnetism, lime mortar with burnt bricks.

Empire State Building-381 m high was built in1931

Progress from 1930 AD till now is clearly seen all over and it is the duty of engineering profession to make living better with much faster space.

In addition to accommodation, new requirements of humans were felt and accordingly engineering has supported by making roads and bridges for transport, high dams on river flood controls and power generation, sea ports for marine transport, air Ports and many more to make living comfortable as much as possible.

Pillars of Ashoka built in third century BC

There has been remarkable development from scratches in fields of following major facilities by respective fields of engineering and education;

- Power generation and its distribution
- Household utilities and comforts
- Hygienic and variety of food
- Civic facilities and administration
- Communication by telephones, computers and television, even by help of manmade satellites

- Road, sea and air transport of men in comforts and their heavy cargo movements
- Defense and many more

Education and employment opportunities have to maintain a sustainable growth in a balance manner according to demand of society with good quality standards. There has to be to be a periodical review of demand of skills and number of seats in each steam of knowledge giving has to be adjusted according to the forecast and only quality professionals should be produced with almost assured a good carrier.

Change of selection of a carrier after passing out from institutions is also due to improper teaching and producing professionals more than required causing unemployment and frustrations.

We would now provide brief introduction to a few civil engineering structures other than buildings.

Before we go further, we would impress that all the humans on this planet should have right to have basic amenities. Heaven and Hell like differentiation between developed and even below the undeveloped countries is pain full.

One of reasons of making this presentation is to spread as far as possible the need to build quality infrastructure in such lagging behind communities and societies and the engineers could at least part them knowledge and advisory support within the means of these societies.

Children of fishermen sitting on shore under a hurt

Fishermen in Papua New Guinea

A native of Papua New Guinea sitting in his hut wearing old and rejected modern clothes from Australia and Singapore,

A house made of forest wood and threshes **still**, common in countries like Papua New Guinea.

Papua New Guinea is a big market for old and rejected clothes, while they are having one of the six richest gold mines in the world and many other precious minerals. Locals have no value for gold and they do not wear gold ornaments nor understand value of these minerals. They still wear ornaments made out of bones and shells.

1.3 Bridges

River bridge building is a major branch of civil engineering and plays an important role in building a Nation, its Economy and Defense

A bridge made of tree roots in Meghalaya

A suspended steel structure bridge on river Hooghly in Calcutta, known as Howrah bridge is shown in photograph below. To facilitate, shipment under the bridge, the whole bridge is suspended by two parallel catenaries of steel beams pieces hinged like a chain one on each side of the road supported by high towers with span of 1,500 ft. (457.2 m) built on shorelines of river and anchored firmly in ground on both sides at a reasonable distance behind the towers. This bridge is of similar design to many old age bridges like one seen in above photograph.

Many such bridges are built in mountains to avoid a support in river bed with high current and big stones rolling with water which would hit and damage piers of the bridge.

In 1992 another bridge was made on this river to ease the traffic between two sides of the river. This is as per modern design of cable stayed bridges built at almost half of the present cost of new steel bridge as per design of Howrah bridge due to advancement of technology.

Road Bridge at Ayodhya, a reinforced concrete bridge built around 1965, much heavier than pre-stressed concrete bridges is still intact as new.

The Mahatma Gandhi Setu Bridge over the river Ganga in Patna is the world's longest river bridge.

1.4 Roads

Grand Trunk Road is the Oldest and Longest Major Road of Asia is in India

This road connects Kabul with Chittagong in 2500 kilometer length and still existing since MAURYA EMPIRE, in 16th century regularly upgraded and until 1980, vehicles and people were carried on boats and barges to cross rivers

Roads have been and will continue to be the important and major resources for communication. Without roads, life would practically come to a standstill.

GRAND TRUNK ROAD

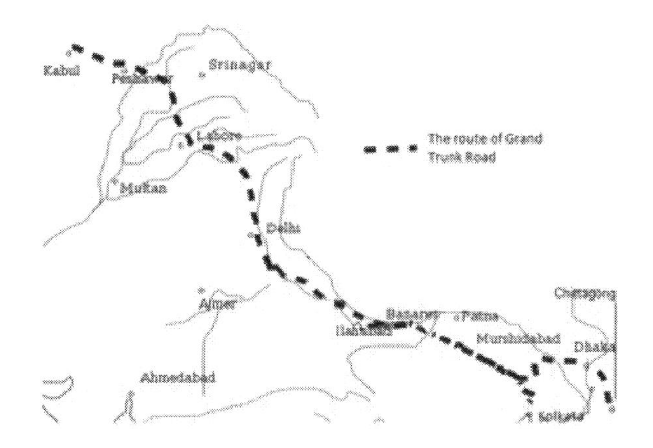

Route Information

Length: 2,500 km (1,600 mi)

East End: **Chittagong, now Bangla Desh**

West End: **Kabul, Afghanistan**

A **Few Details and Photographs of Modern Roads in India**

Being mostly flat land and a big developing country, road and rail transport are most popular and needed in India, air travel in public reach is only about thirty years old that too restricted to business travel by executives and emergencies. Still air travel in India is in reach of only well to do people.

Indian road network is one of the biggest in the world totaling to over four million kilometers. There was a pressure on country, post-independence to increase connectivity with economical road construction and roads with about 60 years' good performance with vehicular speeds around 60 Km per hour were built.

Later around year 1995, country decided to repair and upgrade roads in bad condition and in addition building of golden quadrilateral and east-west corridor by upgrading existing roads and diversions to many townships. Now after substantial connectivity being achieved, India is poised to build good quality highways and super-express road ways. A good scope for civil engineers

At the same time many undeveloped or developing countries have a large scope of road building to different standards according to their present status of economics and social development.

BMIC Cloverleaf: The Bangalore-Mysore Infrastructure Corridor or BMIC Cloverleaf interchange is a 4 to 6 lane private tolled expressway to join India's two fastest growing and green cities.

Badarpur Cloverleaf: The Delhi Faridabad Skyway is 4.4 km elevated highways connect Delhi to Faridabad on National Highway 2. The interchange ends after crossing three lane dual elevated highways at Mehrauli Junction, near Delhi border. The much-awaited Badarpur flyover is to solve the problem of traffic jams at the congested for traffic to and from New Delhi.

1.5 Dams in India

One of the highest arch dams in Asia and third tallest dam in India

Bhakra Dam

Located at Bhakra village of Bilaspur, about 13-km upstream from Nangal municipality, it is one of the highest gravity dams in the world. The lake is about 90 km long, covers an area of approximately 168-square km, of which 90% is in Bilaspur and 10% in Una district.

1.6 Sea Ports

Vishakhapatnam Harbor, one of the largest sea ports in India

1.7 Air Ports in India

Cochin International Airport in the southern state of Kerala is one of the world's first airports runs exclusively on solar power.

Advertisement

Chatrapati Shivaji International Airport Mumbai

LEVEL 2

Basic Education

2.1 Improvements and Innovation

Humans have developed their knowledge in engineering for daily needs and comforts step-by-step, right from the Stone Age to date. They still strive to make something new and better. This process will continue forever.

It is a saying that when sugar is added to milk, it becomes sweeter. Similarly, the greater the interest you have in your work, the greater the urge to do still better and to exert more, and spend time to find better ways to do things. This will bring in higher dividends in all fields, and earn the satisfaction and appreciation of others. It ultimately brings respect and monetary gain.

A civil engineer has a wide range of projects to build and work on, in different capacities. Their photographs may seem scary. A simple question comes to the mind of an ordinary person. "How do they build such projects and huge structures?" A civil engineer says, "It is very simple. We have knowledge, understanding and tools to build such structures. In addition, we also have the will power and determination that are the most important tools for us!"

All these projects are based on engineering knowledge gained not in a few years, but as a result of continuous innovative efforts and methods to improve the skills of civil engineers in the past. Now, better tools and improved skills are at our disposal. Definitely, one will do better and better in the times to come. A few such methods and tools will be explained in this book as we proceed further. These tools are continuously improved upon and are replaced by better ones.

2.2 Professional Engineers and Engineering Training

An engineer, in his professional life and career, comes across many difficulties and challenges. Each of these may need new tools and methods to work with, and the one, who, with his commitment and innovative ideas, succeeds to overcome such situations, is always admired and treated as a successful engineer.

2.3 Knowledge in Multiple Skills is Necessary

In this rapidly changing world, requirements, thoughts and tastes keep changing. Therefore, it is important and necessary to be updated with the requirements and developments, and to improve one's knowledge accordingly. Just sticking to one type of work is not desirable. In such circumstances, one should not miss out on any opportunities to do something new. By missing out on such opportunities, one stagnates. In the professional life of a civil engineer, very frequently, something new has to happen. An opportunity is provided by superiors to a young engineer, by taking a risk on themselves, they trust that the young engineer will do the task well. Such opportunities should never be missed or turned down.

After independence, India prioritized agriculture and connectivity. Many road projects were announced. All civil engineers started building roads, bridges and overhead water tanks in towns, municipalities and villages. This became all the more important after the Chinese Aggression in 1962, since road and rail connectivity in the country is integral to defense movements. A new organization called border roads was created by the Army to build roads in the Himalayas. These were built to connect strategic border areas with each other, and with the mainland.

A few years later, the government needed more electricity in hydro and thermal sectors. A little later, around the 1970s, there was a requirement for ports and harbors. Many ports were developed simultaneously during that period. Now, there is a demand to build intelligent high rise buildings, super highways, railways, huge industrial establishments, airports and atomic power stations. All these projects are different in terms of engineering skills. Only those who are successful in mastering these skills can take up the challenge and quickly switch from one type of a project to another in their careers with good opportunities.

The huge secret remains that all these projects need the same basic knowledge of engineering. The applications are different for different projects. Once you understand the basics, implementation in different modes is easy. Therefore, we should not have a negative attitude at work. We shouldn't say that it is not our work, or ask why we should do it. If a supervisor or a friend asks you to work on a new task, in a way, he is giving you an opportunity to learn more. Obliging him by taking such an opportunity will benefit you too. If it is denied, he will give such opportunities to others in the team.

2.4 Mistakes and Difficulties

Nobody is perfect. Nobody has ever been perfect in the past. Everybody makes mistakes. One has to learn from his mistakes and should not repeat them. There should be enough burning of the midnight oil to find and realize why a mistake has happened, and one should find ways to improve oneself, and avoid repeating the mistakes once made.

In spite of doing the job well, due to too many unforeseen reasons, there may be failures in one's designed process and progress on a project. These failures are sometimes inevitable, and one should not lose heart. It is meant to make one stronger and more determined to carry out the task or mission even more strongly and more effectively, especially after such unforeseen reasons are taken care of.

We once seated huge precast concrete foundations twenty-two meters high, weighing about eight thousand tons each in twenty meters' depth of water in the Visakhapatnam Port in 1972. Every time we tried to place the foundation of a floating concrete structure, the structure itself drifted away by a few meters from the designated location. We found that we were doing the operations during spring tide, considering that we would have a greater tidal difference, i.e., in sea water levels, between high tide and low tide. This had to be taken into consideration for our operation. But, we did not realize that the water current was higher when the tidal water level difference was more, when compared to Neap tides, when the difference in the two water levels is less. The high water current used to push the structure out of the place where the foundation was located. We started the process in Neap tides and did not have any problem thereafter.

Let's take the case of a group of small ants climbing up a wall. If their passage on the wall is rubbed with a finger gently, to make the wall smoother and thereby creating a little static energy, the ants will fall down, but will not give up. They will get up again and again. They will find another way, bypassing this obstruction on the same wall and reach their destination.

If a small ant can do it, why shouldn't you try at least? The same logic is applicable to spiders, honey bees and many other tiny creatures, all of them never give up. Humans are the wisest and strongest creatures on the planet. They are not supposed to give up!

2.5 The Basic Requirements to be a Successful Engineer

To be a successful engineer, one must have the following factors:

- A basic understanding of the subject
- The capability to apply his engineering knowledge in the execution of the task assigned, productively
- The capability to look ahead and plan work in advance
- The capability to handle Crisis Management and Damage Control

For this, parents and teachers have to contribute substantially.

2.6 The Role of Parents

Requirements as stated above are common for all professions. These qualities are difficult to inculcate in a child through tutoring alone. A child is often tutored just to get good marks, in order to get admission in a premier college and then, take life easy. Once admitted into a professional course at college, somehow or the other, he graduates and becomes a professional. Most times, this happens without an in depth understanding of the subjects that are needed in the future, in his every day of professional career.

Becoming a successful man is largely attributed to his upbringing. Ultimately, the person becomes one out of the three kinds as under:

- A person who takes his own decisions
- A person for whom someone else takes decisions
- A person who takes decisions for others or on others' behalf.

The mother is the first teacher for a child. Knowingly or unknowingly, she helps the child by showing him how to stand with her support, to walk and to speak. Having achieved all this, how can she stop teaching her child any further? Similarly, a father also has some responsibilities. Today's parents should look back at the treatment and training provided to them by their parents and the outcome of all of it. This would properly evaluate the needs of their child.

Parents should not create differences between the children in the family based on their interests, gender, intelligence and attitudes. They must provide equal opportunity to all the children in the family, to grow.

The child should have independence in framing his opinions and mindset under the guidance of his parents. But, he should not be forced to take decisions for his life against his wishes, especially if he is not going off track. It is the duty of his parents to provide him with all the opportunities he needs in order to learn, which are well within their means.

There was a higher secondary student, way back in around 1956. His father wanted him to become a doctor. But, he did not like the medical profession. Unfortunately, one day, the argument reached the stage where the young boy got up, emptied his pockets on the table and left the house. He first started driving the car of an engineer. He then learned and studied in his spare time. Slowly, he changed his profession to that of a civil engineer by coming up in ranks. Ultimately, he rose to the level of Zonal Director of a big construction company. This young man, with his determination, reached the highest possible level - a rather difficult for a well-qualified civil engineer of that time.

His parents could have avoided the flash point with him. His leaving the house had put the family at risk. If the parents had cooperated, he would have avoided the hardships. He could have made use of that time well for his professional development. He could have done much better in life, ultimately giving credit to his parents.

There was a successful parent. He did not give his son too many worldly comforts, though he enjoyed in his town lived a comfortable life. He, in turn, inculcated a hobby in his son, of making scientific models for his annual exhibitions at school, with a mandate that each year, he should produce a new innovative model under his guidance. School teachers also cooperated and contributed their expert knowledge. They opened the doors of the science laboratories for him to work in, at his own convenience, without missing out on class. His sons were more successful than he was, and continued to enjoy an elevated status in society, being the children of a well-known man who was remembered as the father of these good sons.

It is always desirable and expected that parents should give time to their children, instead of spending money on their children. If we look at world history, the percentage of successful people coming from well-known wealthy families is less than sons and daughters of parents who have struggled in bringing them up. There is a bigger percentage of spoilt children from rich families. Even in large industrial and business houses around the world, the successful growth of business by the second and third generations are often a question mark if the children did not shoulder the hardships right from childhood with their parents.

Parents and schools have big roles in the development of a child. For their mistakes and negligence, the child suffers all its life. This is generally termed as God's wish, or fate, which are untrue, false and unacceptable.

2.7 Role of Schools

Schools have an equal, if not a bigger, role to play in the development of a student. In the school, a student is expected to be given the knowledge that he needs in order to live in society in a civilized manner. The child is further taught the basics and fundamentals of various subjects like language, literature, history, arts, civics and science and such else. As the child grows, he can use this knowledge to develop himself further and solve difficult problems in society and his career. Educational curriculums are designed by experts to take care of such requirements and are revised often, to maintain pace with the developments in knowledge and society.

School teachers should understand the difference between teaching and tutoring. As a routine, they should reflect once in a while to see that they are really teaching the children while keeping themselves updated with the developments in the system.

We met a minister. He proudly introduced himself saying that he was a professor in the university before his joining politics. We asked him a simple question: **"Sir, you had been teaching or tutoring in your life!"** he was baffled and agreed that he was tutoring and not teaching the whole of his life. In fact, he might have never tried to find difference in teaching and tutoring the whole life, with an objective to just make students clear exams with may not be having clear understanding of subjects.

Competitive exams after 12th standard for admission in premier educational institutions, for engineering and medical studies are a recently developed system. It was felt that the standard of teaching and awarding marks in exams was not uniform across the society. Therefore, a system of competitive examinations had been created to judge students uniformly for admissions into premier educational institutions. This system should work very well if it was still implemented in the spirit of its formation.

A few questions come up here. Schools offer basic education. The intent of competitive exams is to select the best out of the lot of educated children. Why is it necessary for students to go to specialized tuition classes with an assurance of getting selected by the desired institution, irrespective of their caliber and performance in school? Do these tutors provide tuitions for just getting admission into a college of your choice, or to provide basic knowledge? Why couldn't these tuitions be provided at school, as an extension of regular teaching? Why are toppers in tuitions not toppers at school exams? If given a chance, would the institution choose a topper from the tuitions or from school? Are the premier educational institutions satisfied with their system of selection?

If so, what is the percentage of real professionals passing out from their institution? What part of them is good? How do they choose to exercise what is taught for their careers, and to help them succeed in their lives as experts in the profession taught by the institution?

If schools do not do their job properly, why do parents are keeping quiet? Are they not bothered? Are they just willing to pay for education and purchase at the prevailing price? Why do they leave the rest to the fate of their child, which is in fact their own responsibility?

With all-round development in the world, engineering skills have also advanced and provide a good platform to young students to start their profession and career in a particular stream of engineering, after elaborate classroom lessons and practical training.

Considerable time and money is spent on each student for this training. It should not go waste. The student must turn out to be a good and successful in his life.

This is true not only for engineers, but also for doctors, lawyers, scientists, accountants, agriculturists and many other high profile professions.

Further, these professional qualifications, especially from a high profile educational institution, provide an elevated status in society. Marriages are arranged in good families, parents also have high expectations of status and financial support, and many students have to pay back hefty bank loans taken for studies.

It is generally late, but not too late to realize this. If one realizes that the employer is doing business and that he himself is doing business for the employer, and if he does not make the desired returns in terms of profit and progress for the employer, why should the employer keep him? Many people, either ignorant or pretending to be ignorant of this basic truth, find fault with the employer or the environment and change jobs. Ultimately, after changing a few jobs, they are content with what they get and leave the rest to fate.

There is another group of students that can afford to go for higher education or change of profession. Ultimately, they land up somewhere good or bad, depending upon how the change has happened.

If someone has made a mistake in choosing the right career, he requires some amount of brainstorming to avoid making a repetition of his earlier mistake. It would be not only a loss of money, but also a big loss in time and self-confidence. Everybody has some talent, need to be explored and progress using and exploiting this talent to full extent.

LEVEL 3

Basic Understandings and the Importance of Studies

A young man puts in a lot of effort and working with a lot of hope lands up in engineering college. This is the time for him to start learning, and to work for a bright and successful professional career.

Now, the stage is almost set for him to become an engineer. Thereafter, he may construct or destroy, knowingly or unknowingly.

This is a serious statement. There are glaring examples of failure of bridges and industrial foundations due to small mistakes on part of engineers. A few of these will be elaborated further when we discuss such topics in detail.

Now, the engineer is out of the rat race. Getting good marks is necessary to get into a premier collage. The time starts now, though late. But, it is never too late to gather in depth knowledge of all the subjects that are taught to him.

In case he does not understand a chapter, an engineering student should go to the teacher at his time and convenience to seek help to clear doubts. One must understand that in case of tuitions, he pays money to the tutor. He provides tuitions for the bare minimum that helps serve the purpose of answering exams. Teachers don't need the money, but actually need to be respected for the imparting of knowledge in detail, for the understanding and satisfaction of the student. Such knowledge does not have a price tag!

On becoming an engineer, good faith and trust are bestowed by society in the individual. He is expected to perform according to expectations. One goes to a doctor and takes the medicine in good faith and trust. The outcome is the fate of the patient. The doctor charges a fee and has hardly anything to lose. This cannot continue for long. Sooner or later, the doctor will be discarded by society, if he does not perform well. In a similar manner, once a qualified engineer starts practicing, everybody starts trusting him. It is for him and his caliber to justify the trust bestowed on him. The medicine given by a doctor can be poison, too, but the patient has taken the medicine, trusting the doctor. Therefore, one needs to have the required knowledge and skills in a subject before a responsibility is bestowed and accepted.

In engineering college student would be first taught basic subjects related with their applications in engineering. This would prepare student to understand and solve complex issues of the engineering and virtually this would be the base for further studies.

The first few subjects taught may be:

- Elementary of mathematics and its applications
- Theory of structures
- Hydraulics
- Theory of electricity and electrical engineering
- Heat energy and applications
- Chemistry
- Engineering Drawing
- Etc.

If a student feels that a particular subject is difficult, for any reason, he should still try to get into the details and get out of that difficulty. If he feels that this subject may not be useful to him in the future, it is wise to go to the teacher and request him for advice. Why should he study this subject? What are the benefits of studying it?

As we proceed further, we will be using the word teacher and not tutor.

Mathematics

Trigonometry

In field surveys at construction sites, with advanced trigonometry applications an engineer can find latitudes and longitudes of his own location. He marks locations of important components of structure on ground as well as at higher levels by using instruments like theodolite, sextant, Distomat and total stations. All needing the, angles and distances to be feed to instruments calculated by trigonometry applications and having fraction of millimeter accuracy in distance and seconds accuracy in angles. For some calculations eight figure log tables are used.

Arithmetic and Algebra

Arithmetic and Algebra are necessities to be a successful engineer. They are used extensively in engineering design and in making intricate calculations. They are also used in addressing the daily needs of a field engineer.

Field engineers have to calculate quantities for scheduling and estimation of materials and other resources, including manpower, construction equipment and certification of work bills of contractors.

In addition, there are always some small design requirements to be handled. Practically, some application of mathematics is a part of the daily life of an engineer, irrespective of whether he is in office or in the field.

Theory of Structures

Theory of structures is another important subject in engineering. It may be a rocket, a car or a concrete pillar of the verandah of your house. Everywhere, a simple exercise has to be conducted to ascertain loads to be taken by structure and required size of structure. The end product has to safe, practical and produced by economical construction or manufacturing process.

Hydraulics

In every household, water and sanitation services are necessary. One would expect reasonably good flow and efficiency of these services. These are the subjects of hydraulics engineering. The study and application of hydraulic science and engineering are required for the design of flow of water and other fluids in various applications. Water, oil and other fluids are conveyed across long distances, running into hundreds of kilometers. This is done under heavy pressure, safely, efficiently and economically with good speed and quantity. It is designed and executed by hydraulics engineers. Even iron ore is conveyed through ragged mountain terrain by pipelines mixed with water in colloidal form, in India and Brazil. The length of a pipeline in India is over 320 kilometers, and is spread over three states of Andhra Pradesh, Orissa and Chhattisgarh. The knowledge of hydraulics is very important in the design of dams, hydro-power generation, reservoirs and other structures dealing with water.

Chemistry

Knowledge of chemistry is widely used in engineering. When water is added to cement, due to a chemical action, heat is generated and concrete is formed. Various types of natural ores like iron, nickel, aluminum and even gold are converted in metal by an engineering process facilitating the chemical actions.

Light, Heat and Electricity

Electricity, heat and light are major sources of power and energy. They play vital roles in everyday life. Therefore, they are very important subjects in engineering. Practically all streams of engineering have applications of these vital resources of power. Thus, their basic knowledge and applications are taught in pre-engineering and engineering classes. They are important in practicing the profession in any streams of engineering. Interestingly, these sources of energy are used in engineering and converted between each other as per requirements. Electricity is generated by heat in thermal power stations and solar power plants. Even an ordinary calculator runs on light energy, using photo-electric cells. Electricity is used to illuminate, heat and cool houses, offices and industries. The energy is convertible from one form to other.

Engineering Drawings

Engineering drawings are different from artistic drawings. They are simple and do not require artistic skills, but to prepare, one needs a sound understanding of engineering. These drawings are representations of concepts. They offer details of a structure in simple diagrammatic ways for a clear understanding. The final outcome of the design is expressed on drawing sheets, as an expression of how the structure would look once built.

LEVEL 4

Detailed Engineering Studies

As a young man gets into designing structures, his training becomes interesting. He is filled with curiosity to solve many riddles in daily life. We will look at interesting examples on different topics and notice how these issues are taken care of by engineering.

Expansion and Contraction Due to Change in Temperature

At home, often during storage, a glass tumbler gets stuck in another. Just dip the glasses in hot water. The outer glass will get heated and expand, while the inner one remains the same size since not heated. With this a gap is created, the glasses get loose and separated.

While you drive on a bridge, you find metallic strips at certain intervals. These are expansion joints at the junction of every two concrete or steel bridge girders. A gap is provided at each junction to accommodate the contraction and expansion of girders due to the change in atmospheric temperatures. Metal strips are provided at the end of each set of girders in order to avoid breakages at the edges of a structure with the heavy movement of traffic. If we do not provide these gaps at joints of the girders, with the increase in temperature, girders will expand. But, they will not be able to move. Additional force should be induced, and that may damage the girder.

Lightning Arresters in Tall Structures

Electric charge is a property of any matter. There are two types of charges, protons carry positive charge and electron carry negative charge, like charges repel each other and unlike charges attracts each other and a balance is created by transfer of charges to make matter neutral. Earth surface is carrying negative charge.

Good conductors of electricity facilitate movement of positive charge to Earth while bad conductors create resistance to this movement, hence likely to get damaged while positive charge is moving to negative, the Earth.

When huge clouds float in air and collide with each other, energy with a positive charge is generated called lightening. Positive and negative charges attract each other. Therefore, the tallest structure in the near vicinity, being closest to the clouds, attracts this positive static energy and transfer to Earth. This damages the structure severely due to resistance from the structure to the flow of energy. Thus, the structure is hit and damaged by lightning. Sometimes, in plain fields with no structures around cattle, trees and men are the tallest. Thus, they die from the lightning and bodies are burnt since they are not good conductors of energy.

To avoid this natural disaster, a safe, quick and smooth passage for the flow of energy to the ground is provided. The structure is fixed with a lightning arrester. A lightning arrester is a thick metallic strip of copper fitted on one outer side of the building. This strip adequately protrudes above the building with a large area in the air so that the static energy hits the strip and not the building. when metallic strip reaches the ground It is embedded at a reasonable distance from the building, with a good electrical conductor, such as charcoal, which is spread around to enable the entire electrical charge to go straight into the ground. The efficient working of a lightning arrester saves buildings from major damages.

If you look at a high voltage power transmission line, positive charge is being transferred from one place to other through a good conductor. It has to be supported at certain intervals by poles or towers embedded in ground. To avoid transfer of positive charge in the conductor to ground through the towers, ceramic discs are provided as insulators between conductor and tower. Each ceramic disc is fully insulated only up to a certain voltage, hence multiple discs are provided, to make it fully safe and avoid transfer of energy at specific voltage. More number of ceramic discs, higher is voltage of power being transferred.

Use of lubricating oil in gear box and engine of your motor bike and cars

The gear box and engine of your vehicle have fast moving parts in confined chambers. These parts generate heat while moving. They need lubricators for easy movement and less friction. Different makes and grades of oils are marketed for these requirements. With continuous heating and cooling, the oil becomes thick and turns blackish. Since it becomes thick, its efficiency for lubrication reduces and it needs to be changed. This is developed with a combined study of many subjects

These were a few examples to describe the usage of engineering in our everyday life.

It is in the interest of the student to study all that is taught in classes with good concentration and a curiosity as to how and why this happens or necessary. A good engineering student will keep visualizing the behavior of the structure while studying. He will also start solving problems with his knowledge gained on the subject. He should not hesitate to ask the teacher questions to clear his doubts.

Looking at the history of the construction industry, you would have noticed and appreciated that civil engineering is development oriented and new concepts, designs and innovations keep on happening for progress in the society and mitigation of new challenges while implementing new designs.

LEVEL 5

Practical Trainings

Once the engineering student starts understanding his subjects, he will be sent for practical training for a few weeks to an office or establishment by the college. Parents are inclined to use their resources to arrange the training at a location close to their houses. Sometimes, it is even arranged as eyewash, for the certificate alone. This has to be avoided. The training coordinator should have a fair discussion with all students and distribute training programs according to the aptitude of the students. Some organizations pay the students, but regardless, students have to work and produce desired outputs to the training provider. It is good to earn while working, but work should provide an opportunity to learn. That is more important. Sometimes, training without pay is more productive than training with payment.

It is important that if a trainee has a problem in understanding a method, he must not hesitate to ask the training supervisor, or his boss, if his training supervisor is unable to provide him a satisfactory solution. For this, the trainee has to go to senior with an appointment or go when the trainer can accommodate him. Trainees are required to write a training report after the training period ends. It would be a better learning experience if he keeps sharing notes with the supervisor and improve it as per the suggestions so given.

On returning to college, the trainee should give his teacher the opportunity to read the training report and discuss it with him. He must see how to improve his studies and training opportunities further.

LEVEL 6

Project Management

Now, we will move onto discussing practical issues with their applications for the management of a project for a civil engineer. We will not discuss theory. Our discussion will be restricted to applications; otherwise, it will become bulky and unmanageable.

6.1 Health, Safety and Environment (HSE)

With all-round development in world including developing countries compliance of **HSE** requirements has become an important issue in project management. Whatever may be efficient the contractor, he is required to maintain records of HSE compliances and this carries a good weightage in decision making on award of future contracts to the contractor.

Good HSE compliances reduce labor unrest, better productivity, low accident rates. It reduces in project costs by getting more effective man hours, less insurance premiums and reduction of medical expenses.

Environment protection is now a global issue and all over the world more and more stringent rules are coming in force and implemented. To date it is very difficult to just cut a tree in one's own compound, drive a vehicle emitting black smoke or driving a truck full of mud without covering with tarpaulin on top of Earth in the truck.

Health

It is the responsibility of the contractor to take care of health of his workforce working in an environment deterrent to health. At first he should provide suitable means to eliminate or reduce impact of such deterrents and then in addition provide medicines for protection from illness. In old days' employers used to give jaggery to eat after days' work to workers working in dusty atmosphere to avoid dust inhaled in day going to lungs and goes to stomach while eating jaggery along with. Doctors wear hand gloves and cover mouth and nose with a mask while treating infected patients. In towns having high level of smoke pollution, traffic police now wear a face mask.

Following precautionary measures are to be taken related to general health care of employees as applicable;

- Provide adequate clean toilet facilities for workmen
- Arrange proper drainage in case of frequent rains to avoid mosquitoes and other insects brewing at site
- Safe treated water for drinking and cooking
- Hygienic food preparation and serving to work force
- Safe eye wear to be provided according to working environments
- Face masts to be used while working in dusty atmosphere
- Clean beds with good air circulation for sleeping
- Periodical medical checkups for everybody working on the project
- Any other precaution advised by the Doctor

On an airport construction project in dense rain forests, deadly mosquitoes and frequent malaria illness was a matter of routine and big concern. The client was worried of getting a proper contractor, who would take risks on his workforce. Work force was above 2000. Client gave a contract to Specialist Company to cordon off the site and camp areas from mosquitoes by spraying mosquito repellant chemicals regularly before mobilization of workforce and continued thereafter. He also made it mandatory for everybody to take an anti-malaria pill daily. In three years working, there was zero-man hour time loss due to malaria. The total cost was less than treatment cost of malaria for one percent i.e. twenty worker's treatment a year. All the contractors and their workmen were saved from this hazardous environment and finished the project satisfactorily in time without any fear.

Such situations are often faced in some way or other on green field projects at the time of start and should not be taken lightly else would create big administrative problem and many would have personal health and safety problems.

Safety

Safety precautions and requirements on the project site, camp and surrounding society would be explained to new joiners. They will provide a safety kit, and safety manual for the project. On a normal project site, the safety officer should insist on the following as a bare minimum:

- Use of safety shoes and helmets all the time at the work place.
- Proper dress that is safe for work. Generally, a uniform is provided.
- Safety gloves of an adequate standard and a suitable quality while doing physical work, including climbing ladders. Separate gloves are available for different applications.
- The usage of safety goggles or suitable eye wear if working in a dusty environment, or in the bright sun/light.
- Stay away from welding and gas cutting unless you are wearing protective gear and special eye wear.
- Wear safety belts as mandatory and always use them while working at heights and risky situations ensuring they are always clamped and fixed on, stable and firm object.
- Check scaffoldings and scantlings are on stable, flat level base and erected with all joints tightened and fixed as per design and drawing. You should avoid overloading temporary as well as permanent structures. In normal circumstances loading to design load needs elaborate checks to avoid accidents. 80 to 85% of design working load is safe depending on quality of structure to be loaded.
- Barricade the working area of cranes, excavators and similar equipment and prohibit unauthorized workmen and officers entering the barricaded area while equipment is in operation. Generally, officers get hurt more often due to overconfidence.
- Check the safe capacity and condition of lifting tackles and the crane before lifting a heavy item of being of safe capacity and not worn out or damaged.
- All cranes have safe working capacity according to condition of the crane. Further the capacity reduces with increase in working radius of the crane and height of the object to be lifted. More is the height;

more would be length of crane boom and lesser load lifting capacity. Lifting capacity reduces with increase in length of crane boom and working radius. On greater heights, wind pressure is also often a big consideration.

- The crane has to stand on stable, firm and level ground having enough bearing capacity of ground to take care of the loads of crane and item to be lifted. A layer of hard wood sleepers says 250 mm × 300 mm × 2 mts long under crane pads would distribute load to bigger area and stabilize the crane. Further the crane should never stand on muddy soil and operate. In case of mud, remove the mud, fill up with gravel, compact and then position the crane on this platform and avoid further in grace of water by providing a drain.
- Check that the crane is safe to lift in all the positions of the crane. Capacities of cranes differ with different positions of load and more precaution has to be taken, when crane has to swing with load slowly.
- Frequently if load is lifted in front of pads and swing with high speed it topples with load, since it was not safe for loading on sides of pads and while swinging at high speed load moves a bit outward increasing working radius.
- Keep the site clean as far as possible and good approaches.
- Keep the project site adequately barricaded to avoid onlookers entering the site and get hurt in state of anxiety and ignorance.
- On one of building project, initially there were more than two hundred spectators on daily basis standing outside the compound for hours since this project was something new. Contractor erected a strong, see through fencing all around, one could see happenings on the other side and posted a few security guards to spread safety awareness and hence no problem.

There are many more precautions to be taken depending on the type of work and its specific requirements. We will study a few of them when as we discuss such operations and these precautions are to be highlighted in the safety manual.

The safety officer of the contractor is responsible and monitors safety compliance and advises on the safety precautions to be taken. It is always preferred to have a knowledgeable safety officer at site. The safety officer along with his team gets involved while doing any work having potential risk of accidents to the structure or workers. They first give a brief of safe working methods and then ensures that work is being carried out according to safe practices and instructions. His instructions to stop work cannot be overruled even by Project Manager and if Project Manager tries, safety officer will mark a copy of his instructions to state labor department and get absolved of his responsibilities.

If a floor slab at 45th floor of a high rise building is to be fixed, he would first check that all the workers are physically fit to work at that height, since many get nervous on looking down even from little heights. It is a mental state of all humans that they can scale only up to a particular height safely and this differs widely from person to person.

In addition, he approves designs of the safety system for safe working including falling of material on ground which may injure a person walking by. All over the world one can see top few floors of a building under construction are covered on outside by galvanized iron sheets and safety nets spread all around the building.

State Government labor commissioner and his managers also make periodical visits to all sites and have authority to stop work if found unsafe working.

In case of accidents and deaths of workmen, government pays for treatment and compensation through employees' state insurance corporation (ESIC). In India an act to this effect was enacted by the Indian parliament in 1948. ESIC charges fixed sum from company every month worked out on basis of number of employees, their salary and age of all employees and nature of risks involved. As on date in India, only workmen are covered under this scheme and have opened their own hospitals in country at many locations. In developed world even staff and directors are covered by medical insurance sponsored and monitored by the government.

Further there is a misconception that expenditure on safety is a waste. The engineer has two alternatives.

Alternative 1

- Pay very heavy insurance premiums. Past performance and management are factors to decide insurance premium.
- Maintain a full-fledged dispensary at site or give a service contract to a private hospital.
- Damage to equipment and injuries to the work force due to accidents.
- Pay a few workmen compensations with top up Ex-Garcia payments in case of serious injuries.
- Loss of man hours.

Alternative 2

- Pay minimum insurance premium.
- Efficient and safe working.
- Minimal loss of man hours on the project.

Alternative 2 is always cheaper and is being enforced strictly all over the world including developing countries

Environment

Care of environment has now become a global concern and is now filtering down to common man even in undeveloped countries. They plant a few trees before cutting a tree. Basic ideology is that your acts should not be deterrent to others and society. Everybody understands that your work may be important but still could be done without or least hurting others.

Common environment friendly acts are like,

- Not to smoke in presence of nonsmokers. They would have to perforce inhale smoke released by you.
- Not to cut trees, they convert carbon di oxide into oxygen and give fruits and shelter in open sun.
- Use of proper toilets to avoid spreading of germs.
- Avoid forming of water ponds, good for brewing of mosquitoes.
- Etc. etc.

On construction sites one can further contribute to environment protection by following;

- Sprinkle water on construction roads and work area as often as required to reduce dust formation.
- Properly cover trucks carrying loose Earth or other construction material to avoid spilling over on road and dust creation. A small piece of pebble like stone could smash wind screen of passing by vehicle in high speed and cause accident.
- Avoid cutting trees and grow many trees if cutting away is unavoidable.
- Make proper and hygienic sewage disposal.
- Keep the work place and living camp clean and hygienic.
- Etc. etc.

6.2 Planning and Scheduling

On a project, an engineer is given charge of a section of the project to execute. He should first plan his work and then monitor his planning regularly, taking corrective steps as necessary. On large projects, a separate department headed by a planning engineer is created. They monitor the planning and progress of work of all section engineers and report to the Project Manager with a consolidated report. This report is deliberated in progress reviews on a weekly or a fortnightly basis, which is also an evaluation process of performance of all the section engineers and sometimes, on repeated non-performance, adverse decisions are also taken.

Project planning should include the following;

1. Study the job briefly and find out the scope of work.
2. Go to the project site and see the physical condition.
3. If it is barren or has thick vegetation, understand the surface – see if it is level, dry, swampy or hilly.
4. How the surface looks like, soft muddy, sandy or rocky, if rocky which is type of rock i.e., igneous, sedimentary or metamorphic.

5. Does it need blasting for digging foundations?
6. What are the logistic challenges like road connectivity distance from camp and office, local habitation etc.?
7. Any other details you think or advised to explore.
8. Spend some time in office on the documents, discussions with superiors how and when the job is to be executed.
9. Time available for execution of job including time required for mobilization of resources?
10. What are the resources available handy and challenges to mobilize additional resources i.e. workers, supervisors, engineers, technicians and equipment operators, construction materials, construction equipment, construction and working drawings, safety requirement, housing, market for daily needs drinking water and electricity and their reliability?
11. Is it mandatory to carry out some soil or other tests as per contract?
12. With this most of the information is available and side by side it will be nice, if someone has worked out approximate quantities of work like earthwork, piling, concrete of a grade and type, reinforcement bars steel structures, embedment's, fittings and fixtures etc. with their specifications.
13. Now make a fair assessment of what and how work could be done with the available resources and how long to get additional resources.
14. Here comes a little bit of innovative engineering to work out, how safely, as per quality standards in the contract, fast and economically could this project be executed with maximum utilization of local materials, manpower, equipment and other resources.
15. Divide the whole time available in the sections of, time to start first activity, what all could be done with available resources and sequence of start of activities. Balance to wait for new resources, possibility of overlapping different activities and time for commissioning and hand over.
16. Prepare the first draft note on method of construction, also called construction methodology.

17. Prepare the project time schedule covering all the activities and time to be spent for each activity.
18. Work out resources required for each activity and duration of their requirement.
19. Draw a graph of quantity against time of major resources separately for each resource; generally, it would have many ups and downs. Now you need to adjust the time for various activities for optimum utilization of all the resources, called balancing of the schedule.
20. Work out the cost of the resources needed to complete the project.
21. Final trimming of method statement and time schedule, submit for approval and plan execution as approved by superiors.
22. It is further more important to monitor the progress of work and costs. Compare actuals with planning and carry out necessary changes and modifications as work progress in project schedule and resources as necessary.
23. Your planning should also not be too much aggressive; moderately aggressive is always preferred, by keeping margins in delays in activities.
24. Project team should not be kept on too much pressure to avoid fatigue.

Sample Project Planning

Let us take for example planning of a 50 mtr × 50 mtr building seven floors plus ground floor and basement, a building of nine floors to be built in a very remote place where import of any material, equipment or person takes three months minimum. Only cement, re-bars, sand and stone chips and some construction equipment are available locally. Doors, windows, tiling material, sanitation, electrical and elevators etc. are to be imported.

No other constrains envisaged.

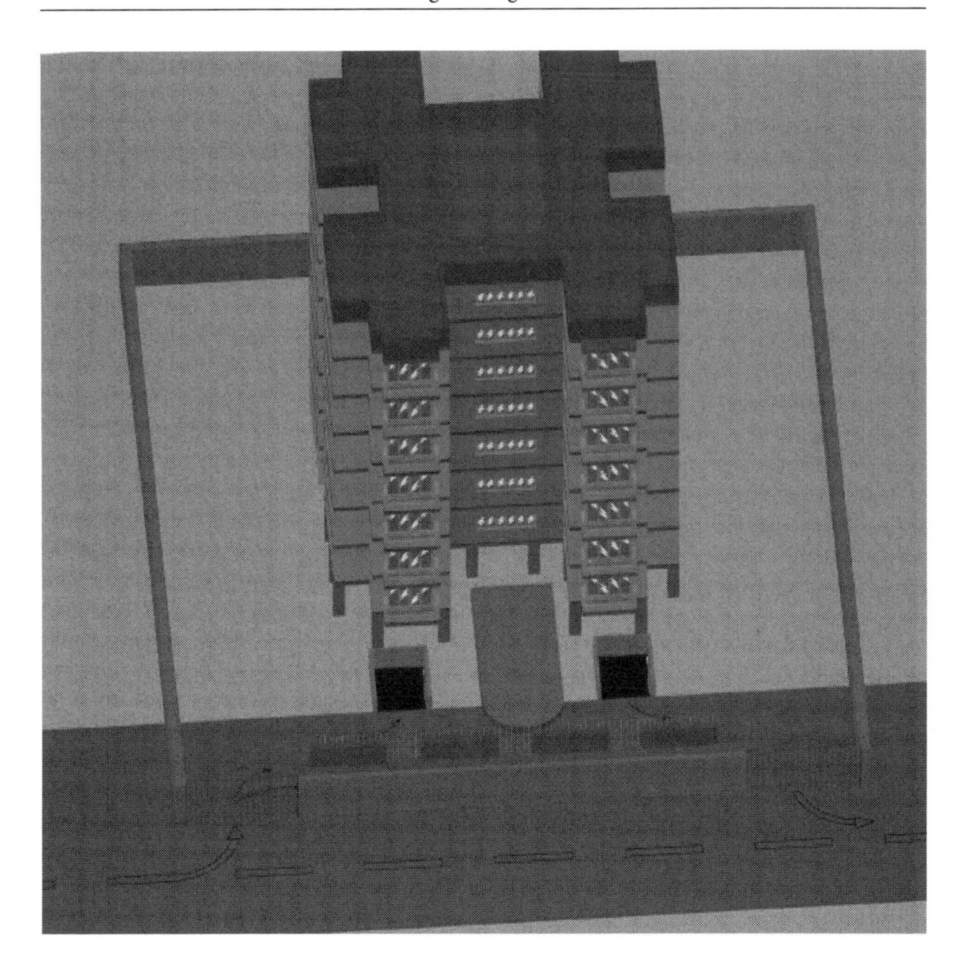

With the availability of sand, aggregates, cement, re-bars and related construction equipment, excavation for foundations and their concreting could be completed. However, one would need some formwork, staging, scantling and equipment to carry concrete to higher levels. One can start excavation with an available excavators or even by man power immediately. Carry out foundation concreting with available formwork. Backfill the foundations and even do retaining walls of basement and flooring of car park. By this time, ply wood for timber formwork, scantling and scaffolding material should arrive, which could have been ordered immediately on getting the contract along with some skilled carpenters and riggers. A few weeks later crane or concrete pump could reach site along with concrete block making equipment.

It is not a matter of laugh, till very recently, 1970 in India there were no excavators in reach of small and medium size projects but still projects were executed. Sheep foot roller for compaction of soils is innovated looking the compaction of soil very effectively while sheep's and donkeys walking on embankment under construction.

A big international contractor in India in 1960's was an earthwork contractor using donkey labor for construction of high embankments.

On a two years project the complete structure including 75% block work should finish in 12 to 13 months and at first stage resources are to be provided locally for Site Manager to complete this much work as per schedule.

This entire planning should be completed in one month to two months. Now the work has started and some cash inflow starts pouring. Planning engineer should now plan out the mobilization of expert man power for finishing works with HR Manager and finishing items and equipment to be supplied for the project in coordination with the procurement executives. Time schedule is prepared in tabular form for the building is appended below. One would notice that many activities overlap. The next activity starts before the previous one ends. It is always advantageous to start any activity as soon as possible., even before it is planned. This gives more time to this activity and a support to the effort of finishing project in time. There are many unforeseen circumstances which always need extra time to attend. A few activities by starting early and finishing early provide big relief in such circumstances. Management appreciate engineers who have contributed in some way for early or timely completion of the project.

Still an efficient planning engineer would provide enough time much more than required for works to be done towards end to cover-up any slippages in earlier activates.

Procurement is also very important activity and should be always well ahead of requirement of men available at site. We need to draw separate activities on planning schedule for all the items to be procured in micro planning.

A list of procurement items is given to procurement department with schedule but generally they procure items first which does not have big hassles and claim so many percent of procurement is finished in less than a month. Delay in procurement of earlier needed make project suffer and early procurement of other items is a strain to cash flow. This should be controlled by the planning engineer. As a matter of fact, before a purchase order is released it should be approved in writing by planning and costing engineers that purchase is being done at the right price and deliveries in the right time and sequence.

On this basis he works out detailed master time schedule for the project which includes progress of work and availability of resources, generally broken down to weekly progress. As the work proceeds and details of actual productivity and constrains are known, he graduates this schedule to micro planning including increase or reduction of resources if necessary. Micro planning schedule should not be done hurriedly without fully understanding productivity, problems and constrains. This micro planning schedule should be monitored as frequently as necessary and is a very good tool to forecast problems, progress, and revenue & expenses flow with constrains, with a sincere effort to complete the job in time.

Micro planning is further breakdown of activities in sub-activities. Each sub-activity has a time slot, which is monitored with actuals and total sum of time required or to be used for these sub-activities is the time allocated for the activity. As the work proceeds, various constrains and benefits are known and recorded. Slippages and early completion of certain sub-activities and overall time schedule is updated periodically.

Preliminary time schedule for a building project of two years' duration is appended below.

Building Project
Planning of Works-Preliminary

S. No	Activity	Months																				
		1	2	3	4	5	6	7	8	9	10	11	12	13	14	15	16	17	18	19	20	21
1	Award of Contract	■																				
2	Setting up Concrete Batching Plant			■																		
3	Construction Equipment Mobilization	■	■	■	■																	
4	Manpower Mobilization	■	■	■	■							■	■	■	■	■	■	■				
5	Start Excavation		■																			
6	Basement Foundation Raft		■	■	■																	
7	Ground Floor Slab				■	■																
8	First Floor					■	■															
9	Second Floor						■	■														
10	Third Floor							■	■													
11	Fourth Floor								■	■												
12	Fifth Floor									■	■											
13	Sixth Floor										■	■										
14	Seventh Floor											■	■									
15	Roof												■	■								
16	Concrete Block Walls								■	■	■	■	■	■								
17	Electrical & Plumbing							■	■	■	■	■	■	■	■	■						
18	Doors & Windows										■	■	■	■	■	■	■	■	■			
19	Tiling, Sanitation, etc.											■	■	■	■	■	■	■	■			
20	Finishes																■	■	■	■		

6.3 Costing and Tendering

Costing is the forecast of costs likely to be incurred in a project till the completion. It s very important for tendering, project implementation and monitoring. If the cost is not properly worked out at the time of tendering, losing the tender due to high price is preferred instead of getting the tender at a very low price. That is not only a loss to the organization, but also creates a lot of problems in cash flow management during execution. This causes delays which further increase the costs due to overstay in the project. Generally, a liberal management would to the maximum provide just to complete the job, free construction equipment and forego profits, unless compelled to put in cash but in such case, would not give a second opportunity to this group of executives. Most contractors take advance payment at the beginning of the job and provide bank credits to run the project at cost of the project. Thus, it is very important that the project starts earning from the first month itself and planning and pricing has to be done accordingly. A good Project Manager should know the principle of "how to fry the fish with its own oil." If one puts fish on a frying pan and heat slowly, fish would release oil and keep on frying as oil keeps on releasing with the speed complimented by both activities. same way project should be executed.

A construction project is generally divided into the following cost components

- Materials
- Equipment to be procured and installed in the project
- Construction equipment
- Temporary and ancillary works
- Safety
- Human resources
- Project management and supervision
- Overheads
- Insurances
- Finance costs

- Risks and their mitigation
- Head office overheads
- Any other costs
- Contingencies
- Profit before tax

6.3.1 Materials

Major materials in civil engineering projects are

- Cement
- Sand
- Stone chips
- Rock in different sizes
- Embankment fills
- Bitumen in different grades
- Steel reinforcement in different grades and sizes
- Structural steel and plates
- Timber
- Building materials including electrical items, doors, windows, ironmongery, sanitary ware, plumbing and drainage material, air conditioning ducting, hardware and other items as required in the contract
- Any specific materials that may be required but are not included in above list.

Cost of Each Unit of Quantity of Materials at Site Would Have Following Components

- Purchase price including delivery to site with taxes etc.
- Any processing work to be included in costs
- Royalties etc. payable
- Loading, transport and unloading at site

- Total unit cost–sum total of above unit costs
- Quantities are available before start of costing for execution of works for each type of material, preferably worked down to monthly requirements. Engineer is advised to fill up the unit costs and monthly quantities in a table for each type of material

6.3.1.1 Cement

Cement is costly and perishable. It has a defined shelf life of about sixty days. Cement gets spoilt when it comes in contact of moisture. By contact with water, it reacts chemically and turns into hard lumps, losing its quality and ability to provide strength when mixed with water. Thus, cement storage and handling are important for project planning, costing and execution. The cost of infrastructure for cement storage and handling should be taken as part of the costs incurred in the project. Wastage for cement should be considered between two to five percent if it is stored in silos or in covered warehouses, respectively. Both are expensive and their capacities and sizes should be assessed properly. A system should be introduced in warehousing on a first come, first use basis for cement.

6.3.1.2 Sand for Concreting, Brick Work, Plastering etc.

Sand is one of most difficult material to access actual costs due to excessive water and silt, both of which are not good for construction. Sand that is purchased and brought to the site should have adequate provisions made in the purchase orders for proper accounting for quality of sand and quantity of water, silt and other impurities present in the supply. If sand is procured on a volume basis, there is a good chance of mistakes creeping in. The volume of sand rises from its minimum volume, even up to 40%, depending on water content that is generally termed as 'bulkage.' Sand has the maximum density when it is saturated with water. It is not practical to saturate sand with water for each truckload, and to waste water every time the sand is to be measured. Thus, if it is purchased on a weight basis, a wastage element of 20% is generally safe for sand costing purposes. 5% for silt, 10% for water and 5% for actual wastage in handling are considered. Sand, unless very small in quantity, should not be purchased by volume. In that case, the actual quantity available would be less than half the purchased quantity.

		US dollars	d	1st month		2nd month		3rd month		4th month		Last month	
MATERIAL	UNIT	COST	e	QTY.	AMT	QTY.	AMT	QTY.	AMT	QTY	AMT	QTY.	AMT
SAND	TON	70	l	3000	210000	4000	280000	4000	280000	2000	140000	0	0
WASHING	TON	3	I	3000	9000	4000	12000	4000	12000	2000	6000	0	0
STORAGE	MTH.	250		1	250	1	250	1	250	1	250	1	250
HANDLING	MTH.	2000	t	1	2000	1	2000	1	2000	1	2000	1	2000
TOTAL COST			e		221250		294250		294250		148250		2250
WASTAGE	20%		t		44250		58850		58850		29650		nil
TOTAL COST			h		265500		353100		353100		177900		2250
NETT UNIT COST			is		88.50		88.28		88.28		88.95		

A good sand supplier would screen the sand through a 12 mm mesh and wash it thoroughly in the river itself, to remove all the silt and clay. This is beneficial for the buyer and seller. Rivers have an abundance of water. If sand is screened before transportation to site, the project would save the transport costs of unwanted material. Further, it will help in saving on the cost of disposal of the rejected material.

6.3.1.3 Stone Chips

Stone chips for concrete are generally known as course aggregates. They are available in two types. One is known as gravel. It comes out of the river in all sizes. The other is from stone quarries in mountains. It is extracted by excavation or by blasting the rock, and then putting them into stone crushers of different sizes. Aggregates are separated based on sizes by the crusher. They are then passed through a screen in sizes of 40 mm, 20 mm, 15 mm and 7 mm. This material does not have bulkage, but the weight depends upon the water in aggregates. For good quality concrete, aggregates in different sizes are preferred to be purchased and mixed on site in proportion, as per laboratory tests. The costing of stone chips is done in a manner similar to the way sand is cost, with a difference that no bulkage and wastage up to 5 to 10% is considered.

For good quality concrete, aggregates should satisfy parameters as required for compressive strength, abrasive resistance, flakiness and chemical impurities and the like. A loader and a few trucks are required as per the quantity to be handled. Sometimes, a bull dozer or a motor grader or both are also added. Foe storage of aggregates a hard stand area is generally prepared by spreading a thick layer of river gravel, with adequate drainage as part of ancillaries and maintained properly. For high quality concrete, the area for storage will have a layer of about 40 mm of concrete grade M15 over hard stand area for storage of sand and aggregates to avoid unwanted material getting mixed in concrete, which will spoil the quality. This helps maintain better drainage and silt removal.

6.3.1.4 Costs of Other Building Materials are to be Worked Out on a Similar Basis

All materials have a fairly constant monthly fixed cost when it comes to their storage and handling. Resources for these activities like storage, handling and security, have to be properly assessed for their optimum utilization. Work progress will suffer if these resources are not provided slightly more than adequate but never less than sufficient.

6.3.2 Equipment to be Procured and Installed on the Project

Equipment like elevators, power standby generators, water treatment plant etc. to be procured, and installed on project constitute another important cost for the project. One is likely to commit mistakes while handling them. For some equipment, only specifications are provided and for some, even the supplier is identified in the tender. Generally, at the time of tendering, a few quotes are collected hurriedly and passed to the estimator. The estimator must check for the following:

- Does the offer for the equipment satisfy technical requirements?
- What is the country of origin?
- Is there a possibility of getting it from a cheaper source?
- What are the logistical benefits and constrains? compare between the offers received?
- What are the delivery schedules? Is there after-sales service? Does it come with warranties?

Sometimes, it is cheaper to provide a discount to the owner if the contractor can provide a cheaper alternative without compromising on basic requirements. This should create room for additional savings to the contractor. On one project tender document, building elevators were specified to be of a particular brand. Naturally, the supplier of that brand increased the price and made terms difficult for the contractor. The contractor did a little research and provided better elevators at a cheaper price to the owner, while making additional money for himself.

In another case, an estimator took one of the major items from a branded company at three times the cost of alternative and he lost the contract. The cost of equipment to be supplied and fixed on the project will have the following components:

- Basic cost of equipment
- Visit to a factory or supplier's warehouse and approvals before placing an order.
- Pre-delivery inspection
- Transport of equipment from the factory to the site including inland transport to the sea port, shipment and the transport from port to site
- Transit insurance
- Unloading and warehousing
- Port clearance and customs duty if payable
- Erection at site, including the erection of equipment like cranes, manpower, foundation bolts, bearing pads etc.
- Cost of supplier's representative for the erection and commissioning of the project
- Fuel and lubricants top ups before commissioning
- Commissioning expenses
- Operation in the guarantee period
- Small repairs and touch up for damages in handling
- Any other expenses not considered above

6.3.3 Construction Equipment

Construction equipment and their operation costs sometimes contribute up to 30% of the total project cost in larger projects like dams and ports. Yet, many times, the project runs in problems due to a wrong choice or the under-estimation of equipment. Hence first a proper list of equipment with numbers and capacities of each type of equipment is prepared in consultation with planning engineer and mechanical engineer in charge for operation of equipment. Invariably mistakes are made in calculation of capacity and numbers of equipment required. Number of equipment decided on add hock basis, need to be avoided if possible.

A concrete batching plant of 60 cum per hour capacity would produce concrete one cubic meter per minute if all the system works as per specifications, including water at defined pressure in required quantity and time added as per manual. There is time loss in positioning of the concrete truck, after one goes off with load and occasionally toping up of material in hoppers etc. delays production. Hence efficiency is about 85% of capacity is generally obtained if there is uninterrupted supply of concrete. But again after completing one job, it has to wait for second job to be ready for pouring of concrete. System once stopped for more than 30 minutes or having worked for long has to be stopped to clean to avoid loose concrete becoming hard inside the plant., hence overall efficiency further goes down to 60% of working capacity of 60 cum per hour i.e., 36 cum per hour, these things are to be seriously considered in planning and often not considered due to lack of knowledge. Similarly, such considerations are to be made for all the equipment on the project.

Mobilization of Construction Equipment is Generally Divided into the Following Categories for a Project

- Own equipment that is available locally
- Own equipment that is available on other projects with logistical constrains
- Hired equipment
- Purchase of pre-owned equipment
- Purchase of new equipment for the project

Own Equipment that is Available Locally

For every project equipment in desired sizes and capacities is chosen with a consideration of optimum utilization and minimum operation costs. More the number of equipment, the more will be operators and maintenance costs, hence for big volumes of work, bigger sizes of equipment is chosen and smaller equipment for smaller jobs. After finishing a job while moving to a new job in the same location, efforts should be made to use the existing equipment instead of getting a new equipment even with some compromises in method of construction. If you have an excavator having reach of three meters, while you need excavator of seven meters' reach for digging up to 5.5 meters' depth, you could dig in two layers. If concrete requirement is 250 cubic meters of concrete per day, you could still do it by working average 14 hours a day instead of 8 hours a day by working in two shifts with a 30 cubic meter per hour capacity batching plant (Working at 60% efficiency) if available at site, instead of getting one of 60 cubic meter capacity from other projects. Such adjustments are very much required.

Own Equipment that is Available on Other Projects With Some Logistical Constrains

Equipment available in house elsewhere even with some logistical constrains should be tried to move to this project instead of purchasing a new equipment. This improves cash flow of the project and company gets some return from such equipment instead of idling.

Hired Equipment

Hired equipment is always a big money drain if taken on daily or hourly hire basis. Many times hiring agency recovers cost of equipment from the job and still owns the equipment. Ones on a project in remote area, the Project Manager decided to take tents on rental for staff accommodation and office for six months, expecting regular camp to be ready by then. But due to unavoidable circumstances, could not release the hired tents for eighteen months. The supplier recovered one and half times of cost of new camp by supplying used camp and still retained ownership. Such incidents are very common and wasteful due to wrong decisions of project management team. World over hiring of trucks and cranes is a big money drain. While the equipment is hired for eight hours' usage, somehow circumstances would warrant overtime working increasing bill value of hire charges, out of control of the Project Manager. If hiring is unavoidable, equipment should be hired on measurable quantity of work done and not on time charter basis. Like an excavator is hired on the basis of quantity of Earth it digs in the month and is paid on measurement of quantity say one Dollar per cubic meter.

Crane is hired for round-the-clock availability and paid for actual hours used with a token minimum monthly billing.

Purchase of Pre-Owned Equipment

Pre-owned equipment is often a good bargain but it requires good skillful inspections and maintenance. A contractor purchased two motor graders, worked for just two years at half the purchase price of new and then used them for over ten years without problem. In another purchase deal contractor purchases an underwater excavator for maintenance a sea port, without going much in technical details. The excavator did not work for a single day since sea bed conditions were not suitable and ultimately scraped. Hence one has to be very careful in buying second hand equipment and it becomes more difficult if equipment is purchased to work in a remote place and does not work properly. However, this happens to new equipment also.

Purchase of New Construction Equipment for the Project

New equipment should be purchased in reputed brands as per specifications of the project for the requirements left after confirmed availability by options stated above. There may be some special equipment irrespective of their cost have only one time use on the current project and future usage is not visible. Such equipment should be fully charged to the project including their mobilization and de mobilization costs. Still before buying a detailed exercise has to be done for better alternatives and possible changes in method of construction.

Costing of Construction Equipment for the Project

For each type of equipment, get the following details in a tabular format:

1. Name of the equipment with the details of the make, model number and capacity
2. Number of units required
3. Owned, hired or purchased
4. Total purchase price of equipment ex works=Unit cost price x no of units required.
5. Delivery to site and transit insurance

6. Cost of commissioning including visit of service engineer
7. Total cost of purchase (Items 4 + 5 + 6) say **A**
8. Number of days or months required on project say **B** months
9. Rate of depreciation to be charged to the project per day or per month say **C%** per month of total cost of purchase
10. Portion of costs, to be charged to the project = **D** = (A × B × C) / 100
11. Residual value after project is finished = **E** = **A** – **D**
12. Number of operating staff required
13. Cost of operating staff including maintenance staff
14. Cost of fuel required
15. Cost of lubricants required, which is most often 15% of the cost of fuel
16. Cost of spares, generally 15% of the cost of fairly new equipment
17. Insurance, logistics, safety etc.

In certain cases, demobilization cost of equipment is also added to costs to project

Owner of construction company makes an investment in an equipment to work on a project. Project Manager would use the equipment on the project, earn money and return money to the owner in a manner or time as agreed or according to established practices. He cannot return the entire investment from next bill or keep on using the equipment as if it is a gift from the owner of the company.

Some accountants charge delivery to site, transit insurance and Cost of commissioning including visit of service engineer as expenses at site and not include in capital cost of equipment to have immediate income tax benefit by increasing cost to project. But it gives a misleading asset declarations at later stage.

There are two methods for charging to the project termed as depreciation.

Straight line method, in this method if life of project is four years, charge 25% of cost of equipment every year and cost of equipment, still running after four years' cost is zero.

Depreciation charged on written down value in this system, you charge every year on written down value of the equipment. In first year you charge/ write-off 25% of investment. The residual value at end of one year is 75% of cost of equipment and project would be charged 25% on the residual value i.e. 25% of 75% = 18.75%, next year 14.0625% and so on. As the equipment gets old maintenance costs are increased and that partly offsets the difference. With this system cost of equipment would never be zero and owner will always have ownership rights on the equipment.

If the Equipment is Rented to a Contractor

If a set of equipment is given to the contractor for use on the project, terms of given such equipment has to be drafted very carefully and not to accept, contractor's proposal without proper study and scrutiny. Even draft contract made by a junior should not be just okayed.

Obra dam project in UP, India, year1975–80 state government purchased a big fleet of Earth moving equipment from United States and gave to contractor for use on chargeable basis. Contractor had to pay to government at monthly depreciation cost worked out on straight line method i.e., fixed monthly rates for use of equipment. Maintenance and operation was in scope of contractor. By end of project book value of equipment became zero even before the project was completed. Contractor smartly avoided to pay hire charges after costs were fully paid. Through court contractor took away all the equipment for free and did many projects with the same equipment thereafter.

The owner should have taken following precautions;

- Charged depreciation in his books of accounts on reducing value basis instead of straight line method.
- Value of monthly rental should not be equal of depreciation value.
- Word depreciation should not have used in the agreement with contractor.

Contractor argument to the court was that equipment survived for more than life expected out of it since he maintained the equipment well and paid the cost of equipment as depreciation as per terms of contract. This argument was accepted by the court.

Next issue is to decide the life expectancy of the equipment. All the equipment depending upon their ruggedness and usage have different working ages. This age also depends upon how roughly or carefully the equipment is used. Taking a balanced approach and the owner could sell the equipment at its written down value at end of the project depreciation should be worked out. For normal use, yearly depreciation on written down values basis, could be as under. However, these figures are purely arbitrary.

- Heavy duty cranes 8% of the costs per year plus interest costs
- Bulldozers, loaders, graders etc. 12%per year plus interest costs
- Heavy duty trucks above 30 tons' capacity 15% per year plus interest costs
- Truck trailers 20 tons' capacity 15% of costs per year
- Smaller trucks and trailers are to be charged at 25 to 30% of costs per year as per site conditions.
- Jeeps, busses, pickup etc. to be charged at 35% of the costs
- Executive cars and jeeps are to be charged at 30 to 35% of the costs per year. Generally, these vehicles are replaced after two to three years and old vehicles are given to site for general use.
- Pumps, small generators, tools and tackles are to be fully charged to the project.
- In all cases total cost charged to project for any equipment should not be less than 30% of the costs of the particular equipment unless project is very small and further utilization is committed.

6.3.4 Temporary and Ancillary Works

Temporary and ancillary works for a project should be planned, designed and quantified properly. Generally, some or all such works are not paid for. Thus, the cost has to be included in the workings of the project. Such works for a building, township or factory are as under:

- Project boundary and specified internal areas fencing for proper security and safety
- Haul roads

- Temporary drainage
- Hard stand areas for storage of materials
- Hard stand areas for heavy crane bases for their safety
- Field office
- Material stores
- Temporary power and water for project execution and their testing before commissioning
- Dewatering system of adequate design and efficiency.
- Toilets and connections to nearby sewer lines or the provision of a septic tank at site, while in some cases client insists on portable sewage treatment plant
- Clients office with committed facilities

For roads and highway, in addition to many of above:

- Temporary diversions of an acceptable standard
- Temporary traffic signs and their monitoring round-the-clock with power connection
- Make shift offices, stores and area fencing
- Breaking of existing structures
- Identification, protection and diversion of underground and overhead services

For bridges dams and sea ports a few to count:

- River training system
- Cofferdams for foundations
- Temporary jetty for shifting material by boats and barges across the water body
- Special measures and specialized temporary scantlings and shuttering for specific conditions and locations
- Special construction equipment, pre-stressing equipment, launching trusses, well point system, piling gantries, special form work and heavy dewatering pumps etc.

- However more detailing could be done on specific jobs and they differ for each project.

6.3.5 Safety

Major costs for safety precautions and compliance are as under for safety measures discussed earlier.

- Qualified and experienced safety management team on site
- Safety awareness training to all
- Safe clothing for all workers and staff
- Safety helmets and shoes for all
- Safe working areas for heavy lifting equipment properly defined
- Safe scaffoldings and scantling with authentic load test certificates
- Safety nets
- Hygiene, cleanliness and nutritious food
- Health care
- Accident management system with general awareness
- Weekly holidays and safe working hours on working days to avoid fatigue and stress.
- Proper sports and entertainment in free time.

6.3.6 Labor for Construction-Human Resources

Labor force required for construction in divided into following categories:

- Skilled workers like welders, tiling masons, plumbers, electricians, high skilled carpenters, structural steel fabricators, fitters etc.
- Semi-skilled workers like shuttering carpenters, scaffolds, masons for plastering, brick layers and junior workers in the category of skilled trades
- Unskilled labor force to work as helpers and to provide necessary support to skilled and semi-skilled workers
- Equipment operators, equipment maintenance and repair crew is considered in the construction equipment category

- Safety management force including watch and ward at site, camp stores and offices.
- Office assistants, peons, cleaning, store assistants, gardeners, camp maintenance including laundry and toilets Drivers for light vehicles.

Costs of labor force should be calculated in the following manner:

- Work out a histogram of the labor force for the actual productive and direct work on site on the basis of project planning schedule. At costing and planning stage about 25% extra to be considered, since average efficiency of workers is between 75 to 80%
- A number of workers could be hired locally in all categories, not needing accommodation.
- Transport to their normal residence costs to be provided by company, especially in late evenings
- A list of workers sourced from distant locations, in different categories
- Costs for accommodation, food, laundry, mobilization and repatriation, including travel during vacation.
- Labor laws applicable at site.
- Unit rate wages of all the categories on the basis of labor law or as prevailing in market, whichever is higher
- Work out total cost of labor on monthly basis and include in the main costing schedule
- Costs of overtime working, taking at least three hours per day for all workers for planning purposes

6.3.7 Project Management and Supervision Cost

An organizational chart is prepared for the management and supervision of the project. This organization should be in form of a triangle, i.e. with minimum staff on the top and progressively incremental staff towards the lower levels of responsibility. The following things should be taken care of:

- Working conditions are important in deciding levels of competency and number of people required in each category.
- The job profile for each person should be well-defined and be preferably written for each one. He should be responsible for his job.
- Procedures and controls are important considerations while preparing an organizational chart. Many organizations give plenty of powers to line management staff and at the same time keep a vigilant eye by good procedures and controls. In such cases Management Information System (MIS) reporting team has to be strong
- Depending on the job requirements, the skills of the staff in all categories should be decided. It has to be balanced. A more skillful person will be more expensive. At the same time, if a staff is of lesser knowledge, would make more mistakes. His superiors would have to help him, instead of doing their own job and this often goes in chain reaction and senior bosses get overloaded or work suffers.
- Five to seven officers should generally report to a senior officer.
- Twenty-five to thirty workers reporting to a supervisor and foreman are desirable.

An organization chart usually has the following departments and disciplines. It is not necessary that one officer is responsible only for one department. Work has to be divided by giving more departments to an officer or if department is very big, divided the work between two or more officers, finally try to manage uniform work load on everyone in the team.

- Project Manager/project Director
- Planning, scheduling and monitoring progress
- Management Information Service and reporting
- Costing and tendering
- Billing
- Finance
- Accounts

- Cost accountant
- Safety and medical/first aid
- Construction superintendent
- Section engineers
- Junior engineers and foremen
- Supervisors
- Construction equipment operation, maintenance and management
- Procurement
- Stores and warehousing
- Security
- Camp upkeep, maintenance and hygiene
- Time keeping, attendance records, overtime working, and preparation of wage bills
- Administration, personal relations, travel, guests and events management
- Industrial relations, local laws, and enforcement of cordial relationship with society and local public

After drawing the organizational chart and estimating the number of staff members required, a list of requirements of the management is prepared along with details of the job profiles of each of the staff. The staff is requisitioned and recruited or transferred from other projects in a phased manner, in accordance with the schedule. Their costs are to be worked out in the manner similar to labor costs. It is important to note that every time a project is planned, generally, manpower, staff, construction equipment and materials are not procured or arranged in the planned sequence. This always causes setbacks to the project. The procurement and management teams just dump what is available and then take care of the items that are left out - even the ones that reach the project site first are to be chased, and reach the project with difficulty.

ORGANIZATION CHART

PROJECT MANAGER

PLANNING	CONSTRUCTION EQUIPMENT	PROJECT EXECUTION	FINANCE	ADMINISTRATION
SCHEDULING	MECHANICAL ENGINEER	CONSTRUCTION SITE TEAM	ACCOUNTS	CAMP
COSTING	FOREMAN		FINANCE	SAFETY
BILLING	OPERATORS		TAXATION	HR
MONITORING & MIS	MAINTENANCE		COST ACCOUNTANT	IR

6.3.8 Overheads

The following items of costs on a project are generally grouped in project overheads:

- Office establishments including furniture in proper layout
- Office equipment, like drafting machine, drawing printers, computers, printers and photocopiers, vending machines, etc.
- Internet connectivity in office and field offices
- Telephone and fax
- Power and drinking water including their installation
- Toilet facilities
- Reception desk and its management
- Conference room with facilities of overhead projectors and video conferencing systems in a proper layout
- Tours and travel
- Guest house setting up including desired facilities

- Expenses to be incurred on visitors, travel, allowances, boarding and lodging and local conveyance.
- Telephone and internet bills
- Electricity and water bills
- Tea coffee and snacks
- Office rental
- Business licenses and their costs
- Postal and courier charges
- Business development and entertainments
- Stationary
- Flowers
- Maintenance of car park, entrance and garden
- Painting and up keep of office complex
- Repair and maintenance of furniture

Some estimators just consider a percentage of the total costs as overheads. This is often erroneous. Practically, many more overhead items, though expensive and important, are not covered in the above list. Overheads differ from project to project. Instead of just following the list, see which item is applicable and which is not and ensure that nothing is missed out. The cost of each item is difficult to establish on the basis of an empirical formula. It differs widely from project to project. Overhead costs should be controlled, but cannot be avoided. These costs have to be considered adequately in the estimate, debated and approved by a competent authority. Otherwise excessive overhead costs are always a matter of concern by senior management and even senior project management staff are ridiculed.

6.3.9 Insurance

The following forms of insurance policies are mandatory at the time of start of the project, and for duration of the project. They may be extended in case of delay in completion of the project:

- Contractors all risk insurance

- Construction equipment insurance
- Marine and in transit insurance
- Workmen compensation insurance
- Personal risk and accident insurance for senior staff and client's representatives
- Motor vehicle comprehensive insurance
- Third party liability insurance
- Cash in transit insurance
- Professional indemnity insurance
- Insurance against war and mob hostilities (if the situation warrants)

Contractors all Risk Insurance

A brief of the project with its activities, risks in execution, design, safety, earthquakes, natural calamity, washing outs and damages by rains, storms and all other possible risks are to be given to the insurance company with the duration of the project in a comprehensive manner. On the basis of the risks described by the contractor, the insurance company should provide an insurance cover and provide a quotation for the insurance premium to be paid by the contractor annually. Insurance companies also ask for the details of the contractor, designer and client experience in executing this type of projects and their track records. It is in the interest of the contractor and engineer not to hide any risks from insurance company.

Construction Equipment Insurance

Construction equipment is insured on the following bases:

- List of construction equipment and their book values
- Condition of equipment
- Safety compliances
- Working condition and risks involved

Insurance company should be told reasonable risks involved in working of the equipment. These risks are recorded and in case of mishaps insurance claim would be considered within parameters of risks indicated at the time of taking insurance. More are the risks, higher the insurance premium.

A crane working on land will have less risks compared to working in floating condition. Hence if the crane is to be mounted on a barge for some time, insurance company should be informed in writing and they would take extra deferential premium for the period. Please also check insurance cover of the barge.

Marine and in Transit Insurance

This is to cover materials and equipment in transit from one place to other against risks of damages and theft, taken individually for each of the major movements of material and equipment, including new procurements.

Workmen's Compensation Insurance

Every country has rules for the payment of compensation to workers according to their age and salary for losses due to sickness, injuries due to accidents, partial disability and in some stray cases, even deaths. The insurance company takes care of the reimbursement of employees, costs of medical treatment and compensation for losses for partial disability. This insurance is mandatory for all construction contracts.

Personal Risk and Accident Insurance for Senior Staff and Clients' Representatives

For senior officers of the contractor and the representatives of the client working on the project, individual insurance policies are taken by the contractor for specified amounts. This is according to the contractor's policy and employment contract with the individuals concerned. Unlike workmen's compensation insurance, amounts of insurance for these employees are not in line with labor department guidelines and regulations. Insurance values are much higher. They go high in cases where high profile experts and executives visit or work on the project site. In some cases, they are airlifted to specific hospitals in cases of minor sickness at the cost of insurance company. In addition to hefty insurance cover, major contractors provide medical treatment to their workers and staff in the best possible manner for major sicknesses or accidents. A well-qualified doctor is available all the time at the dispensary dedicated for the project.

Motor Vehicle Comprehensive Insurance

This insurance is for the book value of vehicles, and is provided for total loss and a specified number of passengers including the driver. They are all insured for compensations in case an accident occurs, and are insured for amounts as desired by the contractor.

Third Party Liability Insurance

Third party insurance is taken for compensation to a person who is not connected with the work but winds up getting injured in an accident related to the contract work. The value of maximum compensation payable to an individual is decided by the contractor depending on the local conditions. This insurance is mandatory.

Cash in Transit Insurance

This is meant to cover the risk of loss of cash in transit due to any understandable reason. The cashier moves with cash from the bank to the office and then goes to the site and camp for the distribution of salaries in cash, to the junior workers. They do not have bank accounts. Associated risks of losing cash in the process are covered in this insurance. Cover is taken for a maximum value of cash in transit as decided by the project management.

Professional Indemnity Insurance

All the structures in the project are designed by qualified and experienced engineers with a good track record. These designs are checked by a senior design engineer. There is a third level check done by the engineer supervising the project on behalf of the owner. The project engineer who is responsible for the execution on behalf of the contractor is supposed to check the drawings of any mistakes before execution.

The contractor is ultimately responsible for the safe and efficient building of a project. As a safeguard, insurance is taken to indemnify mistakes in the design and execution of the project. This is a very difficult insurance cover to get. Only a few expert insurance companies are authorized to provide such insurance policies. In turn, they ask many questions and grill engineers before giving the insurance cover.

Many a time, the contractor has to change expert engineers to satisfy the needs and demands of the insurance company. The value of professional indemnity insurance is specified in the contract but it is quite high for small and intricate designs - say, about 20% of the value of the project. For big and normal designs, it could be as less as 2%.

Insurance against war and mob hostilities. **It is** generally **not** insisted upon by the client. But, if the situation warrants, the contractor must cover his risks. This is a very expensive insurance cover. It is taken for projects in areas of high risk, or areas that are vulnerable to hostilities.

Sum total of costs of all the insurance policies will be the cost of insurance for the project. In case of a delay in completion of the project, a few of these covers will need to be extended for additional amounts of time at extra cost to the contractor. All these insurance policies would have a condition of a minimum claim amount at the discretion of the contractor. Insurance premium is related to this minimum claim amount.

6.3.10 Finance Costs

Finance costs have the following components:

- Performance bank guarantee, generally 10% of contract value to cover clients risk on proper performance of the contract by the contractor
- Advance payment bank guarantee as per contract, generally 10 to 15% of contract value, a security against advance paid by client.
- Retention money-back guarantee, generally 5% of value of contract.
- Cash credit limit from bank, generally 10% of the value of the contract.
- The letter of credit facility for the purchase of equipment and materials, generally 20% of contract value
- Payment of bank charges for the utilization of the above facilities
- Appointment of statutory auditors and payments due to them
- Foreign exchange fluctuations

For a project at the time of start itself, **Sum total of these risks and commitments provided by insurance company and bank are more than double the value of the contract.** Bank and insurance company take the risk against hypothecation of assets of the contractor and provide above facilities on a nominal charge. Every contractor has limitations of taking a number of projects at a time, depending on his net worth. To keep going in his business, the contractor cannot afford to be a defaulter. Otherwise, he bears the brunt of heavy penalties, and financial institutions may not support him the next time.

6.3.11 Risks and their Mitigation

Contractors generally anticipate the following risks. However, it varies on a case to case basis. Many a time, at the time of the tender preparation and costing, the management decides not to bid for the project. Main considerations in taking the project are:

- Is the project within the level the team's best competencies?
- Are the social conditions around the project site conducive and cooperative?
- That there are not many technical challenges for the execution of the project. Does all known challenges are manageable?
- That construction equipment and other resources are available
- That no hostilities and public or labor unrest are anticipated
- The capacity of the client to pay on a timely basis

After the evaluation of the above considerations and challenges, the contractor decides whether or not to make the bid. If he is going to bid, he decides how much to add as costs to mitigate these risks in price.

6.3.12 Head Office Overheads

Costs incurred in head office expenses including office rents, maintenance, salaries of staff, marketing and tendering are fixed costs. This is divided among all projects in proportion to their value and involvement of the head office in their operations on a day-to-day basis.

6.3.13 Other Costs

A quick, but serious review of all costs considered in the estimate and anything that has been missed out, will be added here instead of revising the whole exercise.

6.3.14 Contingencies

There is a difference in risks and contingencies. Risks are taken by the owner, while contingencies are in the purview of the project Director since he is responsible for the timely execution of the contract. He takes a decision based on the following, and accordingly, a certain amount is added to the costs:

- The contract is at item rate or at a lump sum value.
- If it is a lump sum, work quantities should be worked out.
- Any other considerations and challenges

6.3.15 Profit Before Tax

An anticipated profit is added to the total costs. It is generally a percentage of the sum total of all the above costs. Finally, the price for the tender is decided. it is then put on the bid. The heads of costs are reproduced below. Costs under various heads differ considerably from project to project. This is, therefore, an indicative list.

1. Materials
2. Equipment to be procured and installed in the project
3. Construction equipment
4. Temporary and ancillary works
5. Safety
6. Labor
7. Project management and supervision
8. Overheads
9. Direct costs sum of items 1 to 8 55 to 65%
10. Insurances 2 to 3%

11. Finance costs	2. to 3.5 %
12. Risks and their mitigation	3 to 5%
13. Head office overheads	2 to 4%
14. Any other costs	depends on project
15. Contingencies	8 to 15%
16. Profit before tax	15 to 20 %
17. Total price for the project	100%

In case of delays in the completion of the project, all components of the cost are increased, and profits are eroded. If delays are due to the client, generally, they compensate additional overheads. But the project engineer has to impress upon the client that the effect of delays would be felt overall. It is better to record and advise the client well in advance and repeatedly, about the substantial additional costs that he will incur in case of delays.

Total Quality Management

Total Quality Management (TQM) is a comprehensive and structured approach to execute jobs as per the approved quality and to improve the quality of products and services with continuous feedback. The **International Organization of Standards** provides certifications under the **ISO 9000** series and monitors progress periodically through approved auditors.

The implementation of the TQM system is comfortable for everyone: the client, engineer and the contractor. All issues related to the execution of the project are according to the specifications and are written down. They are then deliberated and approved. After this, the contractor has to execute the job as stated in the quality document and the supervising engineer has to monitor execution according to the quality document. TQM document is prepared as per the guidelines of the ISO 9000. It includes necessary details of the project execution in the prescribed format. A few of these details are listed below:

- Project execution methodology
- Project time schedule at the micro level

- Organization for project management and job profiles of senior executives and their qualifications
- Material management
- Quality control and quality assurance, approvals and related documentation
- Equipment operations and management
- Safety manuals and their compliances
- Documentation in prescribed/approved formats

Implementation of TQM system on projects all over the world is considered an easy and important tool, hence preferred on large and important projects. Even the large construction companies try to standardize their reporting formats as per ISO requirements and get the progress monitored under ISO certification and guidelines. It is a privilege to have ISO certification for the project/company.

Representatives of ISO authorities monitor the progress and records and visit project site as per their requirements and audit as per quality documents.

LEVEL 7

Project Execution - Materials Management

7.1 Earth Work

All projects need to adjust or change the shape of the ground in and around the project site. This is done by digging the Earth, reusing it, throwing away or bringing Earth from the surrounding areas. Ditches, if any, are to be filled up and the ground level should be raised. Earth work is predominantly major and is an important component of all the construction projects and main activity for the following projects.

- General reclamation and area-grading to convert wasteland into useful area for agriculture, industries and townships
- Roads and Highways
- Airports
- Dams

Soil mechanics a subject dealing with properties of rock, its behavior and strength in various situations/circumstances, is an important subject in civil engineering. **Earth** as commonly known and is available all around in different types is also of termed as rock in soil mechanics.

7.1.1 Rock in Different Qualities

Rocks are a major component of Earth. They represent the mass in different forms like clay, sand, solid rock mass on the ground and molten lava at varying deep depths in kilometers in the state of high pressure and heat. Water, gases and vegetation are not rocks. Different types of rocks are formed by transformation of Earth's crest (upper layer of a few hundred meters) by natural processes. They are classified as:

- Igneous rock
- Sedimentary rock
- Metamorphic rock

Igneous rocks were formed by natural cooling of molten lava coming out from volcanoes and spread on ground. This is found in form of big sheets of rock, generally grey in color spread in very thick layer in kilometers of length and width. Sometimes it is also found in form of boulders formed by big drops of lava scattered in a sizable area. These boulders and sheet rock with passage of time get covered at some places by volcanic ash and Earth by natural processes of landslides, conveyance by wind and water etc. Ingenious rock is preferred for most of the construction projects, since it is hard, least abrasive, strong and least affected by many of the chemicals.

Sedimentary rocks were formed by small particles of rock compressed with age, heat and pressure and become solid mass by now. These particles, generally carried by rivers, might have settled somewhere and with land slide etc. deep buried under Earth. Sedimentary rocks are second preference, since it is abrasive and weaker to igneous rock. Sand stone is sedimentary rock.

Metamorphic rocks were formed long ago by natural calamities, major transformation of Earth surface, landslides etc. in which big forests and other natural resources, including igneous and sedimentary rocks were toppled and covered by big land masses. In the passage of time due to heat inside the Earth and big load on top they got converted in the form of rock, kwon as metamorphic rocks. Metamorphic rocks are available in different kinds depending on their chemical composition and formation. These rocks have limited use in construction industry like igneous rocks. Marble used in buildings and other applications. Many metamorphic rocks are very expensive including lime stone, coal, diamonds, iron and other minerals. Rocks with high metal content are called ORE, like iron ore, gold ore etc. are carefully excavated and taken away for further processing. Gold ore looks like mud but has up to 85% metal component of gold, silver, platinum etc.

Based on their physical properties, these rocks are classified into the following categories for building purposes:

- Stones in different sizes
- Non-cohesive soils
- Cohesive soils
- Silt
- Organic soils

7.1.2 Stones in Different Sizes

Stones and boulders as generally known are hard rock pieces in different shapes and sizes are mostly available in river beds. They are also available in hills in fairly large blocks, needing blasting for removal in small and manageable pieces. They are further processed by cleaning, washing and crushing by mechanical means for manufacture of stone chips, concrete aggregates and road making material in shapes and sizes as required.

7.1.3 Non-Cohesive Soils

Non-cohesive soil particles do not stick to each other in their natural condition. It is coarse-grained and does not retain much water in their body. Non-cohesive soils are generally formed after continuous beating and abrasion of rock by water and air. Big boulders fall in rivers in mountains and are carried away by strong currents and are continuously reduced in size as they travel in river with water due to hitting and abrasion. Finally, they become sand and then, silt. Stones and sand are non-cohesive materials.

7.1.4 Silt

Silt is partly non-cohesive soil. It has very fine grained particles and is found with some amount of very fine particles called clay. Their usage in construction depends on the grain sizes and the clay content.

7.1.5 Cohesive Soils

Cohesive soils are fine grained rock particles and are commonly known as clay. Due to their very closely spaced fine grains they have good water retention properties. Generally, soil as one sees openly is cohesive soil.

7.1.6 Organic Soils

Organic soils are good for agriculture and forests. They are seldom used in construction. Generally, they are a mix of clay and organic components.

7.1.7 Selection of Earth for Filling in Roads and High Embankments

Geotechnical properties of an area do not change suddenly and it is misconception to think or plan that suitable Earth would be available nearby, hence design the structure with Earth of specifications other than those available around the site should be avoided. In such cases either there is a compromise with quality of work or job becomes expensive by adopting ways uncommon on construction projects.

Selection of Earth has Generally Following Considerations.

- Earth should be free of organic soils, log stumps and other perishable material. After decay these materials would deteriorate, create loose pockets and effect stability of the embankment (big bund built by filling Earth).
- It should not be susceptible to combustion like high content of Sulfur. These are very expensive minerals found in isolated pockets and, should not be wasted in making earthen embankments. Further they could burn when hit by a strong tool while digging, causing heavy damages to environment, equipment and workers around.
- In places where precious ore is available, generally ore is covered by some overburden or the ore of less content of the precious material not good for economical processing, hence could be wasted. This waste material has to be used for construction of roads and buildings.
- Sulphate content in form of SO3 in the Earth should not exceed 0.5% by mass. It is highly corrosive to steel re-bars and concrete structure. Therefore, if high Sulphate content is present in some isolated pockets of Earth, such pockets should be avoided for earthwork for structures However, if unavoidable, Sulphate resistant cement called SRC cement should be used for construction.

Reinforcement steel bars should also be used with epoxy coating of approved specifications to protect from corrosion due to presence of Sulphate. All sides of concrete foundation should be covered by HDPE sheets and anti-corrosive paints.

- Generally, Earth confirming to specifications is available around the project in borrow areas. Borrow areas are the nearby plots of land/hills from where Earth could be excavated and used for the project. In stray cases Earth as available does not satisfy basic requirements stated below:
  - Liquid limit should not exceed 70% of soil mass is water by weight.
  - Plastic index should not exceed 45, the difference between indices of liquid limit and plastic limit.
  - Free swelling index should not exceed 50% of compacted volume.
  - Density of the soil for embankments up to 3 meters' height not subjected to excessive flooding should not be less than 1.52 tons per cubic meter.
  - Density of soil for embankments exceeding 3-meter height or any heights, subjected to long periods of inundation should be not less than 1.6 tons per cubic meter.
  - Material used for sub-grade, shoulders, verges and backfill, density of fill material should be not less than 1.75 tons per cubic meter.

Approval criteria cited above are functions of water content and grain size distribution of soil. If we could control water content in soil and adjust grain size distribution results would be satisfactory. Physical state of a fine grained soil depends upon amount of water in the soil system. When water is added to a dry soil mass all the soil particles get a coating of water film on the surface finer is the grain size of the soil, more is the surface area and more water is required. If we take an object in any shape. It has a surface area. Now if we cut in two pieces, total surface area of both the pieces together is more than surface area of the object before cutting.

In any soil if water is added it provides a lubricating film on particles which helps compaction by a mechanical means. Right quantity of water to be added for good compaction is called **optimum moisture content.** This is determined in laboratory by compacting the soil with different water contents. A graph is plotted for water content against density and optimum moisture content in soil mass which would have maximum density and would take maximum load on it, is worked out.

If water is further added, the size of water coating on soil particles would be increased, volume would increase, bonding between soil particles would reduce and load carrying capacity is reduced. On further adding of water the soil mass would become soft called plastic stage. The boundary between semisolid and plastic stage is called **plastic limit.** Plastic limit is determined by rolling a small quantity of wet soil with finger on a glass surface like a thick thread of 13mm. as one keeps on rolling, moisture content would slowly reduce and keep rolling till 13mm dia thread is likely to crumble. This is the plastic limit stage. Check moisture content by weighing and heating and percentage of water content in the soil mass is plastic limit.

If water is added to the soil mass further, it reaches the stage of becoming liquid. The water content at the boundary of plastic and liquid stage is called **liquid limit.** There is a small apparatus called Casagrande cup, soil well mixed with water in different proportions are put in the cup of the instrument and cut in two halves by a standard tool of 13.5 mm thickness. the cup is repeatedly dropped 10mm on a hard rubber base and number of blows required to close the gap of 13.5mm in about 3 to 4 cm length is recorded. Results are plotted on graph with different water contents and number of blows to close the gap. Water content required to close gap with 25 blows is considered **liquid limit** of the soil.

Difference between liquid limit and plastic limit is called **plastic index.** A soil compactable to a good density and shear strength at the optimum moisture content and there should be a good reasonable difference in optimum moisture content and plastic limit is suitable for earthworks for roads, high embankments and other structures. More the difference, safer is the structure. Approval of soil parameters as stated above are only guidelines and do not over rule project specifications.

Following measures are to be used to improve suitability of any unsuitable soil with support of laboratory tests and approval of the engineer.

- Drainage
- Sand Dozing
- Lime, fly ash or cement Dozing

7.1.8 Drainage

In construction and maintenance of embankments quick drainage of water and avoid it getting into foundation and embankment is very important. Sub-grade preparation is done by filling and compacting Earth to its optimum density at optimum moisture content. Soil would lose strength of the earthwork if excess water is mixed by any means. Hence while construction and also for providing good maintenance. after completing the job, proper drainage is provided along the embankments for water freely flowing away through good and effective drains on both side at bottom and top surface of embankment while construction and thereafter.

In case of expansive soil is used for embankment and plastic limit is on borderline, it is all the more important to provide drainage and avoid additional water getting mixed with soil in embankment. It is desirable to provide HDPE or such other cheap and effective waterproof lining in drains along the bottom of embankment to make drain sides water protected. Water protection of slopes could be done by boulder pitching with grouting the joints with cement mortar.

There are plentiful of examples of settlement of road surfaces resulting in pot holes and landslides of high embankments due to erosion or week base, all due to bad or no drainage provided on the surface and sides of the structure.

If good impervious organic soil is available a 150 mm to 200mm thick layer of impervious soil with grass turf on slopes could be a good solution for protection of slopes against erosion by rains. In addition, good drains on top of bank are to be provided and erosion of slopes of bank by rain water has to be avoided.

Pot holes are created by water getting to sub-grade soil and make it plastic. In plastic soil sub-base and base course stones would sink and plastic soil would come up filling the voids, making the entire area loose and a depression is created called pot hole. Only repairing on surface does not help. The area has to be dug, repair soil below sub-grade by replacing wet soil with dry soil at optimum moisture content, rebuild the road and repair snags in drainage, the cause of pot hole.

7.1.9 Sand Dozing

Very fine grained soils are difficult to satisfy the acceptable criteria for use in building embankments hence liable to be rejected. In such cases soil grain size distribution could be improved by mixing coarse grain material free of silt and organic impurities with the soil. With this porosity, plastic limit and optimum moisture content of the evenly mixed soil mass would improve and easily satisfy acceptance criteria and ultimately provide a well compacted and stable and strong embankment. These coarse grain material are available locally in form of sand, moorum and stone aggregates. This process is called **sand dozing** to soils.

Mixing of sand, gravel or stone chips between 5 to 20% of volume of Earth would increase porosity and strength of soil/Earth. Maximum size of aggregate mixed with Earth should be not more than 50 mm when used for sub-grade preparation in minimum 200 mm thick layers. However, for embankment construction maximum size of dozing material could go up to 75mm. The percentage of course material to be mixed with Earth and its maximum size of particles should be decided in the field laboratory to achieve maximum dry density and optimum moisture content. Other qualifying criteria, acceptable are checked by making a few trials of routine test of optimum moisture content, density and liquid limit for mixed material. Procedure of sand dozing is simple. The steps are as follows:

- Spread a layer of soil in the embankment under construction
- Spread sand in measured quantity as per laboratory approved test evenly on top of soil
- Plow the area and mix sand and soil.
- Add water by sprinkling to required quantity

- Again plow the mix of soil, sand and water
- Rolling till compaction is achieved

Approval of representative of design engineer is necessary.

7.1.10 Lime, Fly Ash or Cement Dozing

Use of lime, fly ash and cement instead of sand and aggregates yields better results in the construction of roads, sub-grades and embankments. But, in certain parts of the world, it has caused adverse effects in the form of swelling, due to chemical action between soil and such dozing material and roads have failed. Hence, before using these material, laboratory by **chemical tests** should confirm that there would not be any adverse chemical reactions between soil and dozing material. These materials give better results than sand dozing and successfully used in IRAN and some parts of Europe. Procedure for their use is the same as sand dozing. There are many manufacturers of cement and lime dozing equipment and they provide all the technical literature on the subject compatible to their equipment.

7.2 Rock Quarrying and Blasting

Before discussing rock blasting, we must remember Sir Alfred Nobel. He was born in 1833 in Stockholm, Sweden, and was the inverter of dynamite that is used world over as an explosive in weapons. In construction, explosives are used for rock blasting operations. Sir Nobel had 350 patents. He created chemicals for destruction and on a day decided to spent his wealth for peace. A trust was created from his wealth to fund the Noble Prize for peace. Other subjects for prizes were added later. Still industries and laboratories set up by him are the world's best. In civil engineering, mostly, explosives with Ammonium Nitrate as the base are used. Explosive in general use is called gelatin. Gelatin is a safe explosive, available in different grades of strength in 25mmx200 mm sticks. It does not catch fire with a matchstick, but when detonated with a detonator, instantaneously, the whole mass converts from solids to gas at very high pressure. This Ammonium Nitrate gas, in an effort to escape, breaks everything around it. More homogeneous and brittle is the rock, less the quantity of gelatin and bigger is the blast.

Now many more brands of explosives are available in market, but still most of them are with Ammonium Nitrate as base. Detonators are about 8mm dia metallic tubes, 50 to 100 mm long, and closed on one side. At one end of the tube, a pinch of gun powder is kept. It is detonated by a small spark with an electrical cable connected to the exploder, which is a small battery or hand operated dynamo. Detonators are blasted by a small hammer, and damage the area around them. A single detonator will give a basket full of fish if blasted in a pond.

As an alternative to electrical detonators, for small blasting of a few boulders etc. to break them in small pieces' ordinary detonators with a safety fuse are used. They burn at a designated speed. Depending on the time required to move the blaster to a safe place, the length of the safety fuse is decided. One end is inserted in the ordinary detonator and once everything is ready to blast, the blaster ignites the other end of the safety fuse with a cigarette butt and swiftly moves to a safe place before the blast. Cigarette is used only for ignition of safety fuse. Otherwise cigarettes and match box are not allowed in and around handling explosives. It is not safe to ignite safety fuse with a match stick; it may burn more length of fuse which is not desirable.

Detonators and gelatin are never stored together, and kept far away from each other. Explosives are stored in a building of concrete or bricks, at least two kilometers away from the nearest house or shop. These buildings are called magazines having a license to store explosives. Shoes with leather soles and nails, cigarette lighters and match boxes are not allowed in and around these buildings and doors should be kept wide open for air circulation for at least fifteen minutes before entering the magazine. In construction industry explosives are used for breaking rock which could not be broken by normal excavation equipment including a Chesil for foundations, breaking hard rock on a hill and tunneling operations in hard rock etc.

Explosives are also used for excavation of hard rock from hills or pits in different sizes and shapes required for production of concreting aggregates and stones in different sizes on civil engineering projects. The hill or pit from where this rock is removed with or without blasting for construction purposes is called a **quarry**. If the quarry is for minerals, it is generally called **a Mine. The process of** mining/quarrying operations carried out by creating benches on hills or open pits and then cut the benches in a designed pattern for production of rock is called **open cast mining.**

Other types of mining are underground mining and tunneling. Underground mining is used for taking out coal and other minerals physically by men at depths more than about fifty meters below ground and tunneling technology is used for making tunnels for roads, waterways, defense and other similar uses.

7.2.1 Development of a New Quarry for Construction Purposes by Open Cast Mining Method

For construction purposes stones in different sizes varying from five tons to small pieces up to ten grams are needed for construction of ports, buildings, roads etc., everywhere wherever civil construction work is required.

These stones are available in hills full of solid rock mass or stones in small pieces. In case of rock, it has to be broken in big pieces first to take out of hill and then further broken in small pieces. Rock is generally hard and cannot be removed by excavators and hence has to be blasted.

The hill selected for removing rock for construction purposes is called quarry and the process of removing stones from the quarry is called quarrying operation.

Blasting of rock in quarry has to be done in planned manner so that all the rock blasted is easily removed and carried away by trucks. Further at the time of blasting small stones fly all around and could damage properties and hit people around. Therefore, blasting has to be carried out in a controlled manner called controlled blasting and removal of material comfortably after blasting.

For development of a stone quarry, first following checks are necessary

- Inspect the quarry for the availability good quality of rock in sufficient quantities. A few boulders from the area around should be picked up and tested for hardness, grain size distribution and water absorption. It can be done roughly at same place itself. Breaking the boulder by hitting hard by a hammer and the effort required to break indicates its hardness. Grain distribution is seen on the broken surface. Rub the broken face with a sharp edge to study abrasive resistance. Take a measured amount of water in a container put a stone in the container for an hour. Remove the stone gently without spilling the water. Loss of water in the container is the water absorbed by the stone.

- The selection criteria on the above parameters are provided as specifications in the contract.
- See if there is any Earth to be removed from the top of stones, and if so, to what extent.
- Study the depth of water in the surrounding areas to ascertain quarrying in dry conditions.
- Logistically, it is safe to carry out the blasting operations in the area due to proximity to habitation and highways
- Distance from the project site.
- Consent from the land owners.

Once the above are satisfactorily carried out and in principle decision is taken to use this place as quarry,

Start of the job. A few of the following jobs can be done simultaneously. Core sampling and testing in a proper laboratory and area mapping are done for quality and quantity of rock. Consent Agreement with land owners for quarrying operations is taken in writing. Take statuary permissions. Field office and storage of equipment fuel, spares etc.

Clean the face of the hill of all the shrubs. Decide the bench height for rock quarrying. The height of each bench has to be lower than the highest reach of the excavator bucket to be used in this quarry. Work should ideally begin from the top of the hill. Cut the hill in horizontal slices. The bottom of the first bench is marked on the ground. Try to make some areas flat at that level, by bulldozing Earth and chiseling to the extent possible and get a fairly vertical face of the rock. Clean the area of loose Earth and stones. Only after statuary permissions are given in writing for use of explosives on this quarry, start quarrying operations by start of drilling for blasting. Procurement of explosives and their storage in explosive magazine is initiated. All workers and engineer should have previous experience in blasting and are authorized to carry out the operations by the explosive department of government. Explosives department provides license to expert workers, called blasters, authorizing to use explosives for blasting purposes in open cast mines.

Before state of drilling, drilling pattern has to be decided. One has to carry out small calculation for drilling pattern, size of drill hole and calculate the explosive to be charged in each hole with or without delay detonators. An experienced engineer in blasting operations would decide the drilling pattern and quantity of explosives to be used for a blast.

An example of such calculations are shown below, and the drilling pattern and quantity of explosives to be charged are purely arbitrary and author is not responsible in any manner.

Gelatin consumption is between 75 grams to 200 grams per cubic meter of solid rock. This difference is dependent on the brittleness of rock and the strata. If it is good sheet rock without any loose pockets, explosive consumption would be less. But if there is either soft rock or big boulders embedded into the Earth, explosives required would be higher. Gases will get escape routes through week pockets and the energy is wasted.

Distance of hole from rock face is called burden. Spacing is distance between the holes. 25 mm × 200 mm gelatin stick is 100 grams i.e. 0.98344 tons per cubic meter. Diameter of hole to be drilled is known with drill machine at site. Drill hole is loaded only half of its length with explosives and balance is packed hard by dry clay balls called stemming.

If bench height is 8 meters and hole dia is 50 mm, Trail block to be blasted, say is 25 mtrs long and 8 meters wide. If bench height is 8 meters and drill holes dia is 50 mm, for a block of 25 meters × 8 metres, considering 150 grams' explosive per cubic metre, 240 kgs gelatin is required. Each meter of hole would take 2.0 kgs of gelatin. Hence total explosive column required is120 meter and total length of holes required 240 meters for 8meter bench drilling depth is 9 meters as per sketch.

Hence total number of holes required is 27. Would recommend to distribute them in two rows 13 number in front row and 14 number on its back equally distant with burden of 4 meters in front of each row.

In addition to this a few toe holes are often provided to give a lift to the blast and keep the bench clean and level. Secondary blasting is resorted to big boulders with a small charge in a 30 mm dia hole drilled by a hand held machine.

Blasting Operation

After drilling is substantially completed, the engineer calculates the quantity of explosives required and arranges to bring them to the site from the magazine. For each hole, one detonator is required. Detonators commonly have a 6 millisecond delay after the spark. For big benches, blasts are done for multiple rows with half a second delay series. In this case detonators would be exploding after half second, one second, one and half second and so on. Once explosive arrives and drilling is completed, all the equipment including drilling equipment with their accessories are moved away from bench and electricity is disconnected. Blasting crew and all those present should not have match box and cigarette lighter in their pockets.

Now all the holes are charged with explosives by the certified blasting crew by putting one detonator with explosives in each hole slowly and tamping gently by a wooden stick. Aluminum pipe with wood plug at bottom is also allowed. After filling the explosive in calculated quantity, balance of the hole is filled with dry clay balls properly crushed and compacted by the same wooden stick. Detonators should be used in such a sequence that first toe is blasted and gives a lift, then the row immediate next to quarry face and then rows behind one by one, as bread is cut into slices.

Detonator wires are joined together in a series, connecting the detonators. The two ends are pulled together and connected to the exploder at a safe distance. The area is cleared up to safe distances of any person and equipment is parked safely. Once all clear and ready, the blaster triggers the blast by pressing the knob of the exploder and the rock that is loaded with explosives will get blasted and falls down on the bench. As work proceeds further, enough area is leveled on the bench and next bench is started. In the same manner and both the benches proceed together.

Blasting is mandatorily to be carried out only between sun rise and sunset. If blasting is done in sunlight, any onlooker, could see the blasted stone pieces coming and move to safety, which is difficult in night. hence as a rule blasting is not allowed in night however drilling operations can be done round-the-clock without use of explosives, hence safe. In case of accident due to blasting in night punishment to the engineer -in-charge multiplies many folds.

After quarrying, blasted rock is carried to washing and crushing plant for further processing as required for concrete. For other uses, it is segregated at quarry itself by loader and sent directly to site. Rejected material is separated out and used in quarry development works.

BURDEN
SPACING
DRILL HOLES

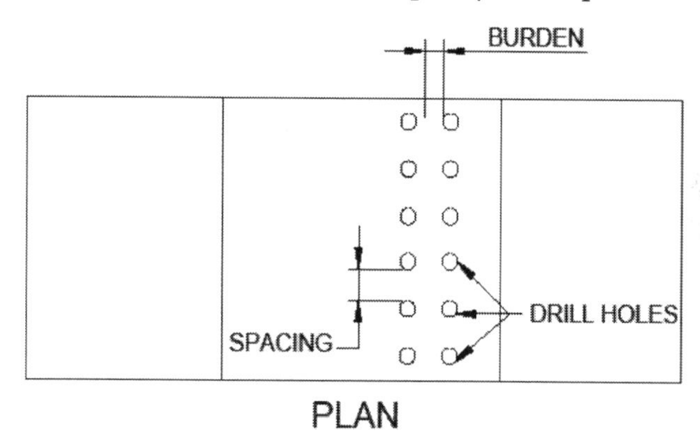

PLAN

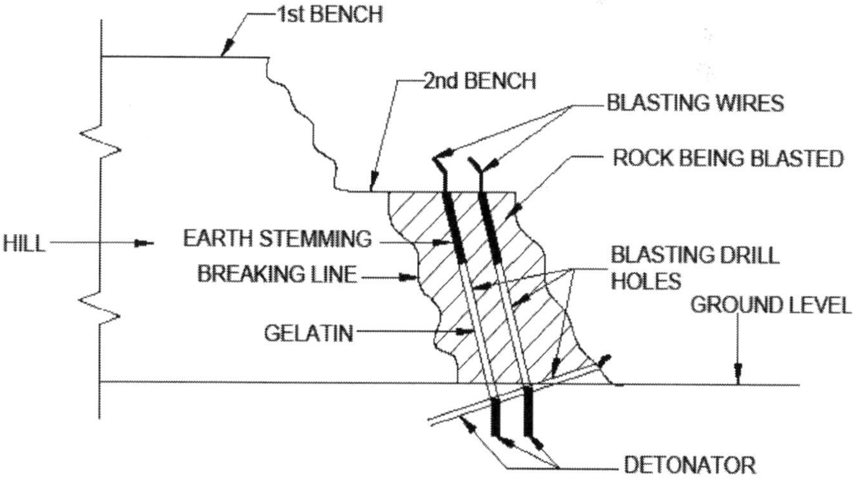

1st BENCH
2nd BENCH
BLASTING WIRES
ROCK BEING BLASTED
HILL
EARTH STEMMING
BREAKING LINE
BLASTING DRILL HOLES
GROUND LEVEL
GELATIN
DETONATOR

CROSS SECTION

Blasted rock from quarry after necessary processing is used for road works, an important ingredient for production of cement concrete for various applications, railways ballast put under the tracks and other such applications.

7.3 Coarse Aggregates

Coarse aggregates are important and major building material used in construction industry. They are used as major ingredient in preparation of concrete, road construction, railways and filter & protection material in earthen dams, besides many other similar applications.

For use in cement concrete production, course aggregates should be non-porous, dense, hard and durable pieces of crushed gravel, and rock, free of clay, silt and organic materials. Presence of flaky stones, alkali and salt reduce the strength of concrete.

For their applications in roads base courses and bitumen concrete, railway and dams their crushing and abrasive strength are more important in addition to above properties.

Coarse aggregates are produced by crushing big stone pieces received from quarry in sizes and gradations as per requirement of the projects.

Crushing strength of concrete largely depends on crushing strength of aggregates, testing is done by crushing one solid cube of 100 mm size in cube testing machine. Poor abrasive strength is, common in sedimentary and metamorphic rocks effect bonding between cement and aggregate pieces.

Material coming from quarry is first washed and then crushed by mechanical, power driven stone crushers. Then passed through screening plant attached to crusher to segregates the crushed stone pieces in different sizes as required.

Many types of crushers are available in the market and mainly their jaws opening should be adjusted properly to get production of proper sizes of aggregates. Crusher jaws are further maintained by frequent welding for wearing out. Flakiness means one side too long, not good for concrete could be avoided by proper maintenance of jaws. Similarly, sizes of sieves of screening plant are to be ordered as per project requirement. Size of crushing plant should be at least double of average daily or monthly requirement on the project worked on the basis of 10 hours a day and 25 days a month working of the crushing plant, due to stoppages for many reasons. Crushing plants due to nature of their job require frequent repairs and at least two hours' maintenance after six to eight hours of continuous working.

These stone pieces/coarse aggregates in different sizes are carried and fed to concrete batching cum mixing plant and are used to make concrete in designed proportions and strength.

Maximum size of aggregates in any concrete depends upon minimum spacing of re-bars and cover to the reinforcement in concrete. Aggregates are used in sizes of,100 mm, 75 mm, 50 mm 25 mm and 12 mm, designated by the maximum dimension of any side of an aggregate.

The photograph shown below is a typical crushing plant. Blasted rock is fed to screen cum washing unit. All fines including clay and dirt is taken out by this screen with washing by a strong water jet. Cleaned and washed stone pieces are fed to crusher the blue unit in photo through a conveyor. Here stones are crushed go to next screen, sorted out in different sizes and oversize goes back to crusher. From screen there are multiple conveyors, one each for each size to stack yard but for simplicity we have shown only one conveyor. Other conveyors also originated from the same screening unit but different screens.

7.4 Sand

Sand, generally known as fine aggregates., Sand should be hard, strong and durable as fine pieces of stones, free of clay, silt and organic material with particle sizes of 10mm to 150 microns, obtained from a river bed or secondary crushing in the quarry. Sand or fine aggregates should have properties that are the same as course aggregates.

Sand is used extensively in making cement concrete as fine aggregate. This is also used as, as water filter for various engineering applications. Sand mixed with cement and water is used as joining material, called mortar for brick and stone masonry, plastering of walls and laying marble floors and many such applications.

7.5 Water

Water for concrete should be free of impurities like alkalis, acid and organic material dissolved in from human and animal wastes. It should be filtered for clay particles and treated for other impurities before storage in a tank or reservoir. If quality is not good, it should go through a reverse osmosis plant of required specifications before use. Water is connected to the concrete batching and mixing plant with a high pressure pump to deliver the measured quantity in a specified time every time preparing the concrete mix. Capacity of pump is an important factor in effective production capacity of batching plant.

7.6 Cement

Cement is a very important ingredient used in the construction industry. A proper mix of coarse aggregates, sand, water and cement makes concrete. Concrete as it is freshly mixed is like mud, and can be placed in molds or a prepared surface as per the requirement. It has to be compacted properly by skilled masons using vibrators and floats. Concrete gets the initial set and becomes hard within 45 minutes to 60 minutes of its production, and attains full design strength in 28 days, and keeps improving strength substantially up to 90 days. It is very important that good quality concrete is produced and laid out properly since its quality would decide the safety and life of the concrete structures. Many structures have failed due to bad concrete.

Cement is produced by the burning a mix of lime stone, clay, sand, gypsum and a few natural materials mixed as per the design specifications of cement. This burnt material is called clinker. Clinker is ground to fine powder in a rotating mill, then packed in waterproof bags and sold in the market.

Main ingredients and their proportions for cement manufacture are as under

Lime stone	72 to 76%
Clay	6 to 10%
Sand	8 to 12%
Iron ore	2 to 4%
Gypsum	4 to 5%

After manufacture, main constituents in cement are Tricalcium silicate and Dicalcium silicate. Their ratio in cement and fineness in grinding decide the grade of the cement. Fresh cement is fully dehydrated and free of water, and has to be protected from moisture before it is put to use. It gets spoilt when it comes in contact with water or moisture.

Heat is generated in the chemical process of setting cement concrete i.e., chemical reaction between water and cement. This heat increases the temperature of concrete and at a stage free water in concrete at that time starts getting converted in water vapor. Water vapor in excess quantity is not desirable. Vapor eats away the water meant for chemical reaction and tries to escape out, which is easy since concrete is not hard at that time resulting in permanent cracks in concrete and even reduction in strength of concrete. This happens more in use of early setting and high strength cement

Cement is available in the following types:

- Ordinary Portland cement grade 33
- Rapid hardening cement
- Ordinary Portland cement grade 43

- Ordinary Portland cement grade 53
- Sulphate resistant cement
- Pozzolana Portland cement

Ordinary Portland cement grade 33 is the base cement. Other grades were developed in the last fifty years by improving this cement. Comparatively, grade 33 gives the lowest strength to concrete. For grade 33 cement the maximum permissible cement content is 540 kgs cement per cubic meter of concrete. The strength of cement increases as the grade number of the cement is increased. The increase in strength of the cement is primarily achieved by adjustment in the ratio of Tricalcium silicate and Daicalcium silicates and finer grinding of cement. For higher strength requirements, one has to use higher grades of cement. A limit of 540 kgs of cement is fairly the same for all kinds of cement, however, we sometimes go up to 650 kgs per cubic meter with higher grades for very special structures. Care should be taken to avoid excess heat generation and cracking of concrete at the surface. Concrete should thus be done at a low atmospheric temperature. Ice is mixed in concrete while mixing the ingredients are cooled by spray of cold water in their bins, if atmospheric temperature is more than 42 degrees centigrade.

Sulphate resistant cement is to be used in cases when the content of Sodium Sulphate and Magnesium Sulphate are excessively present in sand, aggregates and surrounding soil and water. For marine structures, fertilizer plants and agricultural establishments only Sulphate resistant cement is permitted, with a few exceptions where the use of Pozzolana Portland cement with blast furnace slag is allowed. The recommended threshold values are generally Sulphate concentration in excess of 0.2% in soil sub-structure or 0.3% in ground water. Salt in sea water is sodium chloride.

In adverse conditions of high Sulphate or Chloride content in the Earth bituminous paint coating and PVC or like material film cover are provided in addition to use of Sulphate resistant cement to the foundations of structures for the protection of concrete in foundations getting in contact with Sulphate.

Pozzolana Portland cement is available in different grades like OPC cement, with the addition of 8 to 12% of blast furnace slag or brickbats in the cement while manufacturing it. This cement is generally cheaper ex-factory price but the presence of Pozzolana material in cement reduces initial setting speed of concrete thereby slow heat production, however the ultimate strength remains the same as ordinary port land cement.

Setting of Cement and Cement Concrete

Cement in all grades as produced is a mix of totally **De-hydrated** compounds of calcium, silica and iron in compositions and proportions according to blending decided at the time of manufacture., When cement is manufactured, water in any form is not present and the whole mass is totally dehydrated and ground uniformly as per specifications.

When cement is mixed with water, a thick paste of water and cement is formed and it fills all the voids in the aggregate mass. A series of chemical reactions takes place between constituents of cement and water in the paste and these dehydrated compounds in cement get hydrated by reacting with water, these hydrated compounds have a good bonding and thus give cement its strength and slowly solidify, binding all the aggregates, sand and re-bars etc. together by its strong paste. The process of hydration continues for some time say 90 days and as the hydration of constituents of cement continue in chemical reaction with water, concrete continues to gain strength. Ultimate strength of concrete is dependent on the consistency of cement water paste, defined as water cement ratio i.e. ratio of water and cement in the mass of concrete.

There are five more factors which contribute to quality, durability and strength of concrete

- Concrete mass in wet condition should be compacted properly and should have maximum possible density around 2.6 tons per cubic meter.
- Aggregates and sand should be free of clay coating on their surface and all the pours filled in with water to saturation level to avoid drawing water from cement paste, hence properly washed, cleaned and saturated with almost dry surface. If free water is present in

aggregates and sand, it is adjusted in water to be added as per water/ cement ratio while mixing the concrete.

- Crushing strength of aggregates and sand should be higher than crushing strength of concrete.
- Water should be clean, drinking quality, free of clay, organic, acidic, alkaline and Sulphate substances. For Sulphate presence in water or aggregates Sulphate resistant or blast furnace slag cement should be used.
- During chemical reaction between water and cement, heat is generated and care has to be taken for temperature does not increase beyond the limits to allow forming of water vapors. Otherwise water would be drawn from cement paste and concrete would be brittle. If one adds more water at the time of mixing concrete, consistency of cement water paste would be diluted and strength of concrete would reduce.

Special care has to be taken that water vapors are not produced, water vapors start producing at around 50 degrees centigrade hence concrete is kept cool by sprinkling cold water the process is called curing, immediately as initial setting of concrete is started. Curing water also supplements the water lost in evaporation on the surface of concrete.

For heavy foundations and road pavements, Pozzolana Portland Cement, also known as low heat generation cement is necessary. With this cement, quantity of heat generated is the same but spread in long duration, hence damages to structures by excessive vapor formations are avoided.

Following Precautions are Taken

Concrete work is avoided as a rule at atmospheric temperatures higher than 42 degree centigrade and concreting done in nights or ice is added to concrete in place of water.

Cover the concrete surface with wet jute or thick cloth immediately after pouring and keep the cloth wet

Provide extensive curing, surface temperature should not increase beyond about 35 degrees and surface is always wet for ten to fifteen days after concreting.

Use of curing compounds in place of water curing has to be restricted to low strength concretes with maximum 43 grade cement.

If 53 grade cement is used for high cement content/high strength concrete, a water chilling plant is added to concrete mixing plant in hot climates and chilled water around 4 degrees centigrade is mixed in concrete production. This is a cheaper investment compared to use of slow setting/ low heat generation cement on big projects.

7.7 Plasticizers and Retarders

Plasticizers and retarders are special chemicals developed for better efficiency in handling and placing the concrete.

Plasticizers

In some projects, river clean and good shape sand is not available hence crushed sand is used, this makes concrete harsh and cement water paste does not move in body of aggregate mass freely. To make the process smoother, a small doze of specials patented chemicals about 150 to 200 milliliter per cubic meter of concrete are added at the time of mixing concrete. These chemicals are called plasticizers. In addition, for a structure with very congested steel bars, paste cannot be diluted to reach every corner hence plasticizers are added for smooth concreting.

Retarders

In big towns, concrete is conveyed long distances in transit mixers from batching plant to structure needing concrete. Generally dry concrete mix is conveyed in transit mixers and water is added just before concreting. This method should be avoided if possible, since transit mixer drum is always wet and spoils cement in transit. For short distances a small doze of retarders is provided while mixing the concrete. Usually concrete setting time is 45 minutes after manufacture, a doze of retarder extends setting time by 30 to 60 minutes and one gets this extra time to place and compact concrete in its molds.

7.8 Reinforement Steel for Concrete (Re-bars)

Steel is a mixture of iron, nickel, carbon and a few more ingradients. These ingradients are mixed togther and heated at 1400 degree centigrade to make steel in molten state. This steel is molded in shapes of thick slabs. These slabs are used for further processing eother reheating or cold rollong to produce steel in different shapes and sizes, required for different industries including construction. Steel girders, channals, angle, bars and plates are a few.

Steel bars for construction are also produced by melting sceap steel and make small round bars called **Ingots** and rolled in hot condition reducing to required sizes, this is also called secondry produced steel. Since this steel is produced by mixing steel of different grades and compositions, consistant quality is often doubted and questioned

Steel slabs or thicksquare bars about 100mm square by primary producers to produce steel bars and other sections are called **billets** and they conferm fully to desired specifications. Steel produced from Ingots is cheaper than produced from billets.

Steel bars are used in construction to make reinforced cement concrete. In this process a cage of steel bars called re-bars is prepaired as per design requirements, the cage is placed in the mold of the shape of concrete to be produced. Wet concrete is poured in the mold and it fills the mold entirely by filling all the corners and voids in the mold including re-bars.

Any structure due to its own weight in addition to moving loads on it is subject to many types of stress and strains like bending, compressing, pulling, breaking and twisting etc. These stressess and strains, termed as loads effect the structure generally in combination of many forms of loads. In civil engineering, such loads are visualized, quantified and arrangements are made in or around the stucture to safely withstand such and all the other possible loads. This is common for all types of structure. One wears knee cap for comfertable walking since knee has become weak or make a bundle of wooden sticks and tie them to make them stronger while acting toughter etc. etc.

Coming back to cement concrete structures and use of re-bars in concrete. Concrete easily breaks when pulled but acts quite strong to resist any push. While pulling the joints between cement paste and aggregates gives way very fast and seperated, hence the concrete breaks on pulling or bending by a small load.

While designing a concrete structure, a fair analysis of all the types of loads, their quantum, location and direction in the body of the structure are analised and whereever concrete is found weak to withstand these loads, steel in desired shape and quantity is added. This steel is generally in form of round shape in long bars and much stronger to concrete. Hence they are called steel reinforcement bars.

Timber fibers, glass, plastic in different chemical comositions are also used as reinforcement to concrete in some special applications like road surdfacing etc.

Earlier Reinforecement steel bars were only plain steel bars in round shape. Around 1965 in India,cold twisted steel bars with ribs came in the market.These steel bars are produced with a few small ribs along the lenth of the bar. Two ends of bars at room temperature are held at both ends and twisted like a rope.

If one pulls a coir or cotton rope, fibers or threads are individually subjected to pulling called tension however as a group but weaker ones yeild early and in the process the rope breaks on a certain load.

If one twists the same rope before applying load, fibers or threads are more interlocked and provide better and united resistence, taking more load and yeald on higher loads. This difference in re-bars is almost **40% extra** load carrying capacity in cold twisted bars as compared to plain steel bars. This capacity is called tensile strength of steel bars.

Further seel bars are glued with concrete and does not slip in the concrete on load applications but with increase in tensile strength, its capacity to resist slipping in concrete also has to increase. This is provided by providing ribs in steel bars thereby increasing surface area and thus more resistance to slipping.

Another product is in marked called deformed bars(without twist). In these bars additional tensile strength is provided by providing better composition, while slipage is protected by providing deformations like straight ribs. This saves costs and time for extra efforts in cold twisting. Thise bars have straight ribs while cold twisted steel has spiral shape ribs.

Reinforcement steel bars are available in market with different specifications of their composition and manufacturing process including quality standards.

Accordingly there is a big variation in their strength and price. One has to be careful in buying these steel bars and better get samples tested in laboratory before purchase.Common tests are ultimate tensile strength,elongation test, shear strength, bend test and chemical comosition.

Epoxy Coated Steel Reinforcement Bars

Epoxy is a chemical produces in liquid form by different companies according to their patented methods and composition commonly used to fill up gaps in foundations of heavy equipment, joining broaken pieces and protective painting with a thick layer on surfaces likely to get corroded by effect of chemicals. It is highly anti-corrosive, very strong to break and have high compressive strength when becoming solid. generally available in two parts to be mixed in the proportions as advised by manufacturer and applied in a very short time. it would become strong in few minutes. Widely known ARALDITE is epoxy material.

For structures in and around Sulphate-affected areas, Sulphate resistant cement is used for concrete. Similarly, reinforcement bars are coated with a Sulphate resistant epoxy paint. This epoxy paint is applied after cutting and bending/forming are complited and before placing it in molds for concreting. If they are painted before forming in different shapes like circle, octagonal etc, the paint is likely to crack and moisture would enter through cracks, and purpose of painting gets lost, hence epoxy paint is aplied only after forming in shapes as per drawing are complited.

Cover to Re-Bars in Concrete

Re-bars are provided in tensile stress zones where concrete cannot take this much stress of its own. When rebar is subject to tension. There is an elongation in steel bars. Concrete is glued to re-bars, hence can not allow bars to slip and small cracks appear in concrete around steel bars corrosponding to differential elongation in steel bars. Thus, concrete cracks around the rebar once subjected to load. The moisture can enter through these cracks and the rebar is likely to rust. To avoid rusting re-bars are adequately covered by concrete to keep cracks within the concrete and does not extend to the surface of concrete. Re-bars are placed at a certain distance from surface of concrete. This is called cover to re-bars. Cover as measured, in the concrete is the distance between surface of steel and outer surface of concrete structure. Precautions to be taken for while using cover blocks are;

- Use cover blocks of approved shapes and sizes betwween sreel bar and concrete surface to ensure cover is maintened.
- If concrete cover blocks are used, their sides should be rough and prepared by the mix same as of concrete
- Clean the form work, especially the base thoroughly, before start of concrete.
- Check to see that the binding wires are not projected in the cover block area. They should be bent inwards.

7.9 Structural Steel

Use of steel in form of roll sections is very commonly used as building material. it's usage has been always given preference in bridges, buildings, warehouses and industrial sheds or factories. Almost all the old bridges were made of structural steel and still steel bridges are common. The main frame of all vehicles is made of structural steel. steel is produced in form of plates, H sections, C sections, L sections and Z sections. Some special sections are also produced by moldings and forging. Molded sections are made by pouring molten steel in the molds prepared according to shape and sizes required. Forged sections are prepared by contineous hammering a loft of red hot steel to forge it to the required shape and size. Welded and cold rolling sections are now common since they are made in small workshops out of standard plates to the shape and sizes required, always cheaper to standard hot rolled sections. Structural steel is highly corrosive and need protection by aplying anti-corrosive treatments. Steel get corroded by alternate drying and getting wet by water and scales of iron oxide are formed. These scales are to be removed before protective painting by either of the following

- Cleaning with mechanical hard brush or similar highly abrasive tool
- Sand blasting
- Shot blasting

All the above processes are meant to remove rust scales and clean the surface of steel. Immediately upon cleaning, a coat of primer is applied on the steel and then, it is painted as per specifications. This subject is delt in detail in topic on structural steel fabrication.

LEVEL 8

Construction Techniques and their Implementation

8.1 General Reclamation and Area-Grading

All over the world efforts are made to use the waste land, the land mass not good for any normal use due its uneven condition and lots of ups and downs, for development purposes including infrastructure development, industries and housing complexes. General reclamation and area-grading is the work to cut high spots/areas and filling low lying areas and making level ground for useful purposes.

It is always advisable to carry our reclamation works properly in a planned manner. It looks very comfortable to fill up the low areas by town waste material, but if a few persons are kept at filling site, to remove timber, plastic, metal parts and cloth from this fill material, quality of fill material would improve, and these material removed from fill may have some salvage value. Better is the fill, less is foundation costs of buildings and lesser risk of uneven settlements, damaging the structure.

In Oman year 1981, there was a natural heap of big boulders about 30 meters high embedded in yellow clay in a prime location. A contract was given to the lowest bidder to level up the plot for making a car show room and service centre. Contractor brought a few heavy bulldozers and pushed all the boulders from top of hill. Boulders rolled down and settled on the bottom of plot. The process continued till plot became in one level. In the process big voids remained in the fill. The engineer should have insisted in the contract that all the boulders would be broken in small pieces or carted away or even advised the contractor as a variation while work was going on. When the building was made, boulders moved slightly, since were loosely packed resulting settlement in building and cracks. The building was demolished, and cement- sand mortar was injected in voids, before a new structure could be built with a heavy raft foundation and light structure. The responsible engineer was not terminated since he was a good asset to the organization and this time pardoned.

To carry out the filling in a proper manner, following steps should be taken;

- Take the ground profile and plot the levels on a drawing.
- Mark the required final level of fill on the drawing and check if any cutting is also required from higher areas.
- Clear the area of vegetation and top organic soil and store properly on a separate ground from both the areas requiring cutting and filling.
- Check, if ground conditions are wet or dry. There may be some slush in lower areas. Remove the slush and store in a nearby open space for natural drying. And to be used later if it gets sufficiently dry before end of the project execution.
- Start filling in layers evenly spread in thickness of twenty or thirty centimeter from the lowest area and build up as work progresses in uniform levels. Generally, in such works client does not ask for compaction to optimum density. But it always advisable to carry out adequate leveling and compaction.
- As a thumb rule the roller should walk freely after compaction is over without leaving any marks on ground. Roller should just roll.
- Please make sure the lowest areas are filled first and then come up maintaining uniform levels as far as possible
- After all the patches of high land are cut to required levels if still extra Earth is required, it should be brought from other similar high spots.
- All the areas from where fill material is removed should be leveled as directed.
- If the reclamation is for gardening or a park organic soil removed from the plot be spread evenly on top of completed fill to make it suitable for vegetation.

8.2 Ground Reclamation in Marshy Area

Generally, along rivers before their meeting sea, low lying areas of land masses are filled with water and create big wet ponds in the natural process. In this area flow of water in river becomes slow and even it backs up in high tides, therefore all the very fine soil particle still flowing with river water settle down in these big pockets. In addition, if soil in and around pond is organic or fine clay, top layer of soil in the pond absorbs water to its saturation and becomes in liquid state. This also happens with water streams carrying town waste full of organic material, with low areas in their vicinity, good for settlement of organic soil mixed with dirt.

These areas having little water floating on top of organic soil and fine clay in liquid state in a few meters' thickness hence such areas cannot be used for any work in their present physical status. These areas are called swamps or marshy land.

On the contrary if this happens in sandy strata in upstream of **river** and still sand particles flowing in its body, sand would settle down in such pockets of land, while organic and fine particles would still keep on flowing with river. This area is called lake with firm sand base and good source of water for domestic use, lakes are seldom reclaimed for other purposes. however, if some part is to be reclaimed it could be done by normal filling with non-cohesive soils from one side up to a level above water level or as explained here under.

Some lakes get spoilt by discharging town waste in them and they become good for nothing. If one tries to reclaim, where the waste would go. Hence such cases are to be treated on specific case basis with first solution for disposal or treatment of town waste and then reclamation or with combination of these procedures simultaneously.

Construction

Reclamation of marshy land is best done with cohesive soil free of organic components. Cohesive soil, generally available around in such locations is brought to the site with moisture content little less than optimum moisture content and stacked and compacted like a high bund all along the pond on top of hard soil, a few meters away from edge of the pond in as much quantity as possible. After compaction and free of dry lumps, this soil would be having more density than soil in liquid state in pond, hence heavier and cohesive, difficult for water to enter in its body.

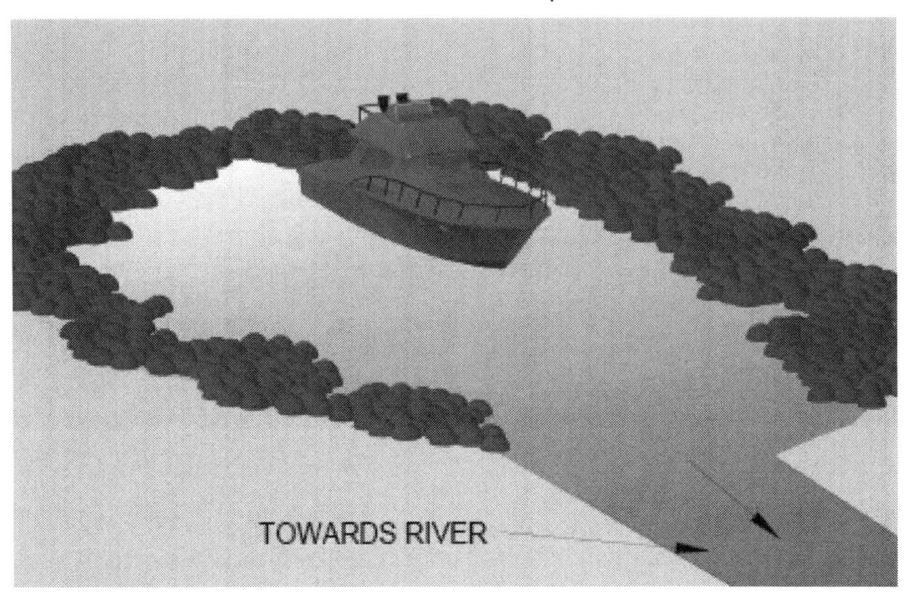

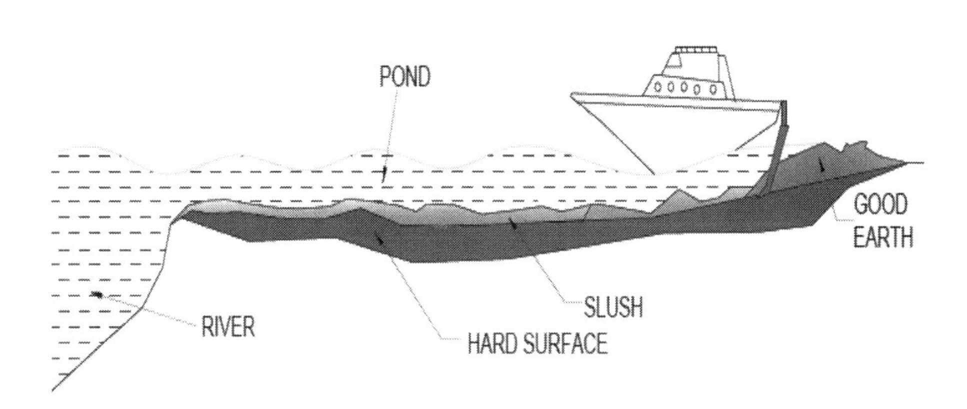

With a few heavy bulldozers, doze this compacted soil very fast from shore end keeping fill level at least one meter above the water level. This dry and high density soil will push the soil in liquid state in pond and take its place, liquid soil would move away and pushing to be done in such a way that it keeps on moving towards river and fell in the river. Care has to be taken that while pushing dry soil it should always keep comfortable exit route for soil in liquid state and never get arrested all around by good soil. It will further help if a few boats are kept to drag loose soil away from pond to the river as shown in above drawing.

Reclamation with boulders is not recommended. Boulders create big voids and loose soil fills up these voids. Soil mass will still retain the plastic properties and low-bearing capacity. boulders would keep on sinking and liquid soil would keep coming up. This is not desirable and is very difficult to repair. However, fill with properly graded mix of stone chips of sizes less than 50mm, sand and traces of silt with practically **zero voids** definitely would be a much better solution. But, it will be expensive and depend on the availability of such material.

8.3 Earth Work for Roads and Highways etc.

For the construction of new roads, or up-gradation of existing roads, a detailed survey is carried out first. Contours are plotted along with desired/ contracted cross-sections of the road, including the right of way. A detailed site visits and survey will provide the following information. Roads and highways have always serious social and logistical issues and the success of the project largely depends on the efficient resolution of these issues. The following things have to be checked at the time of start of work:

- Logistical constrains if any to reach the site with equipment.
- Does any portion of the road have a right of way problem?
- Portions of the road with easy or difficult to resolve right of way issues marked separately.
- Number and locations of buildings and other structures need to be demolished and their status of permits for demolition.
- Existing trees in the road alignment and planning for their replantation.

- Drainage of rain water and water from other sources if any is to be planned and discharge of water is arranged in nearby water body or low areas at reasonable distance to avoid excess water mixing with earthwork.
- Locations of bridges and culverts.
- Portions of the road needed to be dug and portions needing fill and their quantities to be executed in each section up to the foundation i.e. Sub-grade level.
- Suitability of excavated material to fill in the road.
- Borrow areas for fill material with their approaches.
- Portions of the road needing blasting for excavation.
- Quarry location for rock needed for road construction and its suitability.
- Existing traffic conditions in cases of up-gradation of existing roads.
- Social problems if any.

Social problems are easy to solve, if given first preference. Such problems are predominant when, these roads are built to connect habitants living along the road alignment to the main city. Due to low connectivity with town, they are poor and many unemployed. If proper employment, renting of local houses, purchase of local produce as per site requirement and extension of medical facilities, are provided, locals become cooperative and do not create problems. These facilities do not cost more since any way the contractor needs them and by utilizing these facilities saves big logistic cost. Before entering a village for living, foreigners' must make it a point not to indulge in local social, religious and political issues and if any employee is not behaving properly, he should be removed, since sometimes consequences are fatal.

Work planning has to be done for the whole project and resources are mobilized. Work should be started in the area of least problems and problems in other areas to be resolved as the work progresses. A field laboratory for soil and concrete testing should be setup first. Though generally social problems are to be solved by authorities but leaving to them without any cooperation from contractor would delay the project resulting idling of resources and claims before start of project.

8.4 Care to be Taken for Preparation of Road Foundation

If a road is not built properly in a very short time it will be spoilt. the surface would become uneven with cracks. Very soon there will be pot holes, creating lot of discomfort to traffic and even road accidents. A fast moving vehicle will brake suddenly on observing a pot hole and if vehicle behind is not able to control, it would ram into car in front. Main cause for this bad quality is improper foundations. If foundations are not built properly, they would unevenly settle with time and create these problems. For a new road, the construction of a good foundation is very important. This is the foundation of road and if foundation is weak or having week pockets, they would keep on giving problems of uneven settlements and pot holes in the road. Following points may be noted:

- Clear the area of all the vegetation including roots
- Remove all the top organic soil, all wet soil and slush.
- Provide adequate water drainage preferably up to a level of 30 cm below the excavation levels and make it effective before start of further work. In case it is difficult, take foundation work in patches in dry weather and bring the road section by filling and compaction with suitable material above the drainage level in dry period without rains before moving to next patch.
- Compaction has to be done up to density specified for each layer of Earth fill and then go the adjacent patch of road.
- If ground level after removing all the items stated above is uneven, fill up deep pockets and bring them up maintaining level surfaces up to desired final levels of sub-grade and its profile by selected soil.
- Compact the area with a few passes of a vibro roller to attain density of soil at 98% of maximum density at optimum moisture content established in field laboratory.
- On this prepared foundation start Earth filling and build up the road further as per specifications and drawing, maintaining consistent quality and drainage.

Once the fill material is selected and all the preparatory works for road filling are completed, including approaches to barrow area, the work of road filling in embankment starts. Fillings in roads should be done in 20 cm thick layers. Check the water content on the Earth at site and the differences between the optimum water content and the fill material would be the quantity of water to be added per cubic meter of Earth fill. First, spread the Earth in 20 cm layers evenly. Sprinkle water in the required quantity. Leave it for a few hours. Start compaction with a vibro roller till the required density is obtained and confirmed by laboratory tests. Vibro rollers should be 10 to 12 tons' capacity for twenty centimeters layer. Once a layer of fill is approved for the required density, start the second layer and build up the embankment in layers, up to sub-grade level. Please ensure that an adequate amount of drainage is available in case of rains. About 500 mm extra fill has to be carried out on both sides of the embankment. This will be trimmed to the design profile in stages, as the protection of works of sides of the embankment gets started. Drainage and erosion protection works including permanent and protected waterways should not be neglected and should begin as soon as possible.

8.5 Sub-Grade Preparation

Profiles of roads, including slopes and super elevations in curves and bends, are provided in the drawing by the design engineer. Once the contractor is ready in all respects to start the sub-base of the road, final trimming of the road profile is done, as required for sub-grade. Before final trimming, Earth at sub-grade levels has to be tested for moisture and density. Upon approval, the profile is trimmed by a grader and a few passes of a vibratory roller are provided for 95% of maximum dry density.

8.6 Sub-base

Natural material like sand, gravel, moorum and crushed rock or combination of this material are used for sub-base construction. They are mixed to proper gradation and then spread on sub-grade and compacted. The thickness of the sub-base is generally 300 to 400mm and is laid in two layers of 200 mm each. Methods for sourcing these materials are similar to the Earth for embankment. If there are problems getting these materials locally, materials like crushed slag, crushed concrete and over burnt brick bats could also be used with the approval of engineer. The recommended grading for sub-base material with maximum sizes of stone pieces of 75 mm, 53 mm and 26.5 mm are as under. Figures are in percentage by weight passing through different sieves

Sieve size	75 mm size stones	50 mm size stones	26.5 mm size stones
75.00 mm	100	–	–
53.0 mm	80 to 100	100	–
26.5 mm	55 to 90	70 to 100	100
9.5 mm	35 to 65	50 to 80	65 to 95
4.75 mm	25 to 55	40 to 65	50 to 85
2.36 mm	20 to 40	30 to 50	40 to 65
0.425 mm	10 to 25	15 to 25	20 to 35
Minimum CBR values	30	25	20

Strength of Sub-base

The contractor should ensure that the material to be used for sub-base satisfies the CBR requirements as stated above or as directed by the engineer and verified by laboratory tests. Other tests like field dry density, moisture content and such else may be used. Once the sub-base is completed and approved, the base course and surface courses are to be carried out as per requirements of the contract in bitumen, concrete or other base materials. Filling in the depressions in roads, should be carried out in line with the sketch shown below.

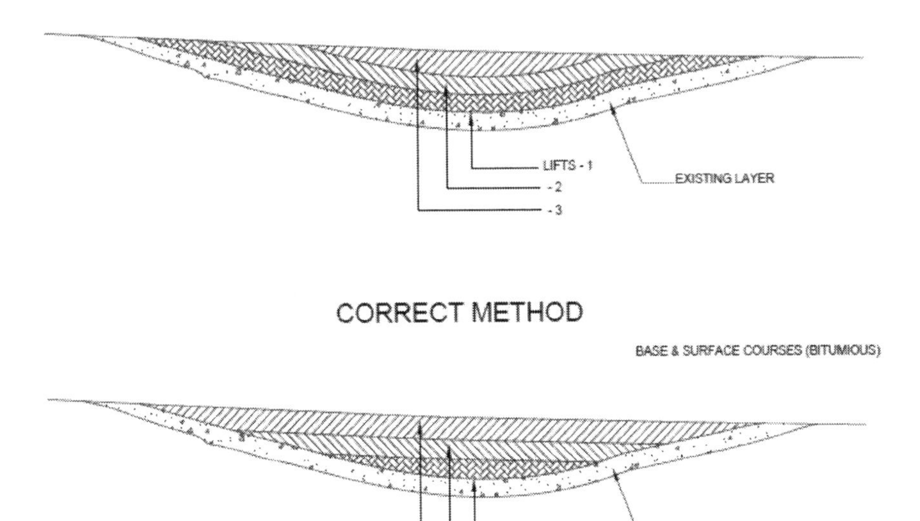

The life and performance of roads largely depend on the foundation up to sub-base top level. But keeping the foundation free of water by providing proper drainage and protection of banks are key challenges for long-lasting performance of all roads.

8.7 Sand Wicks and Stone Columns

A portion of the road or large industrial land may have a very thick layer of fine grained and saturated, highly impervious soil. It is generally black in color and has a spongy feel. It is not advisable to remove such huge quantities of soil. It is not practical, difficult and expensive. The solution is to remove excess water in the soil mass. With this process, the physical properties including the load bearing capacity would improve. This could be achieved by making vertical drains in the entire area very closely spaced and then load the area for some time. These drains would ease the flow of water in the soil which is pressed by load and excess water is drained out. There are two types of vertical drains. Sand wicks are generally 75 mm in diameter. They are driven through the saturated soil in about

1 mtr × 1 mtr grids and are filled with coarse sand. Stone columns are 300 mm to 400 mm diameter columns, about 3 × 3 mtr to 4 mtr × 4 mtr grids filled with gravel or crushed rock, with a maximum stone size of less than 30% of the diameter of the stone column.

For small areas, sand wicks are used. They can be driven by a tripod and hammer powered by a small crane. Stone columns will need proper piling rigs to drive a pipe of a required diameter to a depth of about twenty meters or the bottom of wet soil, whichever is higher. To make a stone column or sand wick, a pipe of a slightly bigger inner diameter than the diameter of the sand wick or a stone column is taken. Now, the pipe is driven in the ground with the help of a crane and a tripod or a piling rig. The bottom of the pipe is closed with a temporary loose fitting plastic cap. The pipe is driven up to the required depth. Since the bottom of the pipe is closed, it would be empty right up to the bottom. Traces of water can come out of the joints at the bottom plug, but not the clay.

Now, prepare the bags a special fabric which allows water to pass through only from one side and does not decay when embedded in Earth, instead provides extra strength to the embankment. This fabric is called geo textile. Bags are prepared by stitching the fabric in the size of diameter of the wick or the stone column of full length of the stone column, keeping the geo textile surface which allows water to pass through from outside to inside. One end of the geo textile should be closed by stitching it up. They will look like pipes of about 20 meters' length, each. Fill these geo textile bags with stones or sand as the case may be. Stitch them close at the top end after it is filled with stones. Once the steel pipe is driven, lift a geo textile bag with the crane hook and lower it into the pipe slowly, up to the bottom of the pipe. It should project about half-a-meter minimum, above the ground after placing in the pipe. Lift the pipe slowly, leaving the bottom plug and stone or sand column in the hole. Go to the next location and repeat the process. An area of a manageable size is selected and positions of the columns are marked on the ground in a desired grid. Work should start in the centre of the plot and precede making columns around, like a circle moving outwards. These columns while driving push the soil away and achieve small compaction. If columns are driven out side to inward they would keep on compacting soil towards centre of the area and ultimately driving of pipe would become difficult and uneven compaction.

Once this is completed, the entire area is evenly loaded with at least half-a-meter of sand and sand wick bags protruding from the ground, are embedded in sand. This sand layer works as a horizontal drain. This is further loaded with good dry soil in one-meter height layers up to a height of minimum of five meters, and kept up to a period of about six months. Cover it with tarpaulin sheets to avoid getting wet by the rain. Make a few level marks on the Earth heap and keep observing settlements. When it is observed that further settlements have stopped, take a sample of the original Earth, at least one meter below the original ground. Test it in the laboratory for moisture content and other parameters as specified in the contract. Spacing the stone columns, the height of the load and duration of the load will be adjusted for further areas as indicated in these tests. For this particular plot, if results are unsatisfactory, the load of stock pile should be increased and kept for longer period under the load, until the required compaction is achieved. To make the task easier, if the permeability test of the undisturbed samples of soil can be conducted, it will help decide the spacing of these vertical drains. The basic exercise is to take out water from the soil with the least effort and time. Excessive and fast-loading is prohibited to avoid the formation and activation of slip circles. This will result in the whole load just sinking and loose soil coming up in heaps around the loaded area.

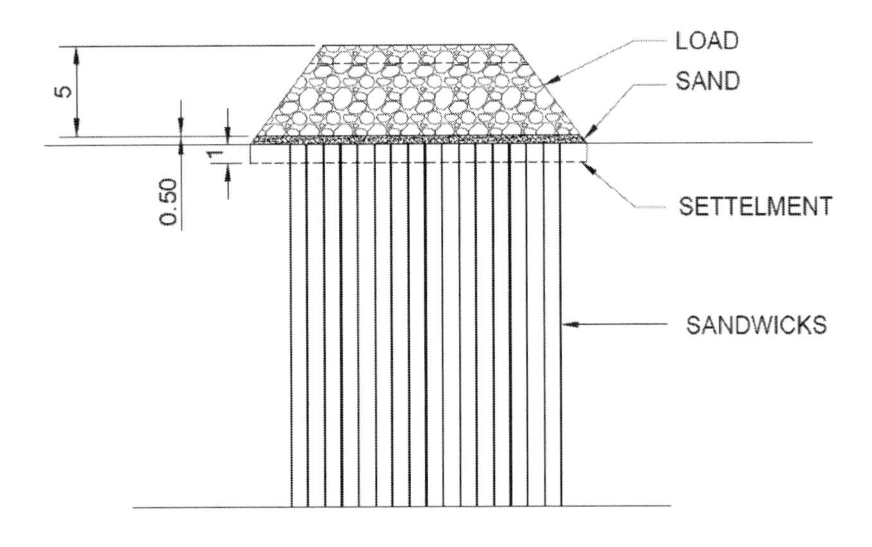

How Does it Work?

This procedure removes excessive water from the soil and increases density by compaction. While driving the pipe in the ground, the soil makes way for the pipe to go down and compresses the soil around it. Thus, it contributes to the compaction, only a little. One would notice water oozing out of the sand wicks around while driving this pipe. Further, once the load is given on the ground, it compresses the ground and releases water in vertical drains around. This water comes up and flows out through the sand layer, which is acting as a big horizontal drain. Slowly, the ground is compressed with a release of water and gains strength. Further, the vertical stone columns or sand wicks installed add to the load bearing capacity of the soil mass, acting as columns.

Still, in some remote areas, long wooden pegs are driven into the ground to increase the bearing capacity of the wet soil. This is an age old process, and has now been modified by sand wicks. In a modified version, plastic vertical drains are also available in the market, and are easy to install. The principle remains the same. In some cases, the ground is just filled with a good thick layer of graded rock, and the road is constructed on this rock fill base. In this method following procedure is adopted:

- First lay a geo textile layer on ground.
- If geo textile is not available, put a thick layer of empty gunny bags or thick wide leaves of plants around, overlapping each other to avoid making the stones sink and clay oozing up to the surface.
- Put a layer of sand
- Now add the boulders
- Again a layer of sand or good Earth to fill voids
- Now build up the road as desired

This method is a temporary measure. Very soon, the road will start having potholes and always needing frequent repair. With time, boulders will sink in clay and clay will come up to the to surface. Like throwing a stone in the mud, the stone sinks. We have to make the mud dry so that we can walk. Instead of stones, if one puts a few jute or plastic bags with small holes, filled with oven dry sand and make a small drain to remove the water, and then fill with Earth it will give better results.

8.8 Cement Concrete

As stated earlier, cement concrete is produced by mixing cement, aggregates and water. Aggregates are inert material in small pieces. There are gaps between these pieces in a heap which are known as voids. Take a measured volume of aggregates in a container and make sure that it is well packed. If the water is added in a measured quantity, it will fill up the voids and replace the air before appearing on the top of fill of aggregates in the container. It will not increase the volume. The quantity of this water filled in voids is the total quantity of air voids replaced by water. Mix half the voids by aggregates smaller to the size of voids and shake the container. These pieces will fill in the voids without increasing the total volume of the container. However, some water that has been replaced by the stones will come out. It should be removed and measured. This volume should be the same as the volume of the smaller aggregates that are added. By repeating this exercise a few times, adding smaller particles of aggregates, say about 50% of the remaining voids, till the particles in the smallest size are added, including sand, without increasing the volume of the container. This volume of stones in the container will ultimately contain the maximum quantity of stones and the least volume of air or water voids. This mass of aggregates is a mix of stones of different sizes and grades, so it is termed as graded volume of aggregates. While doing this exercise, if a few excess pieces remain on top after shaking the container, it should be removed and measured. Simpler methods are provided in books of mix design to achieve the best graded combination of different sizes of aggregates including fine aggregates. This was only an illustration for understanding.

Now, cement and water are mixed to prepare a paste and pour in the container. It will reach all corners and replace the voids. This quantity should be at least the quantity of the voids remaining in the container. Water and cement are mixed in a defined ratio known as water to cement ratio. A table of water to cement ratio for different strengths of concrete is available with the supplier of cement. The lesser the quantity of water in water to cement ratio, the higher the strength of the concrete. The lesser the quantity of the voids, the lesser will be the volume of the water to cement paste.

Thus, the lesser the volume of free voids in graded aggregates, the lesser the paste is required. Thus, the better the grading of aggregates, less water and more strength is achieved for the same quantity of cement. A typical concrete mixing plant is shown below. Big vertical silos for cement storage, circular drum for concrete mixing - and these are fed cement by screw conveyors. Aggregates are fed from behind in different sizes. The truck with a round drum on the back is the concrete truck mixer to carry concrete from plant to site. This truck is known as transit mixer and concrete is continuously agitated by the rotation of drums to avoid hardening in transit.

Concrete Mix Design is Done as Follows.

On the basis of the quality of cement, a graph is available showing 28 days of concrete strength against water cement ratio. Aggregates in maximum and the best grading are also available. Take a few samples of about 250 kgs of graded aggregates in each. Add water and cement in different quantities, keeping the water to cement ratio the same, or slightly higher and lower, in the water to cement ratio. A few mixes will still be dry, and a few will be too wet. With trial and error, get the optimum water and cement quantities in the prescribed ratio. Make a few samples with slightly different aggregate/cement/water combinations. Do not forget to account for the free water in the aggregates and ensure that the correct amount of water is to be added accordingly.

Some of the more important factors are the workability of concrete. Concrete should not be too harsh or too wet to handle. The workability requirements are defined in the design of the structure and are generally measured as slump. There is a conical cylinder of a defined size open from both sides, known as a slump cone. Keep the slump cone vertical on a flat steel plate keeping the wider area at bottom. Fill up the slump cone with wet concrete in three portions, each time tapping the concrete twenty times with a tamping rod. Once the slump cone is filled up to the brim, smoothen the top with a trowel and discard the excess concrete. Lift the cone gently and a heap of concrete in the cone, as it comes out, it will go down and collapse a little. This level difference is measured and is termed as a slump. Now make a few test samples with different cement/aggregate/water ratios and slumps in the form of cubes or cylinders. Test it after three days, seven days and twenty-eight days for crushing strength of concrete in the field laboratory and finalize the single mix which gives it the heaviest unit weight, reasonable slump and the greatest strengths. A few sample test cubes or cylinders are always kept in spare for fifty-two days. Tests should be done, in case of test results are not as good as expected. Test cubes and cylinders are kept wet. They are submerged in water and taken out of the water about an hour before the test, to be reasonably dry at the time of a crushing test.

Besides the water to cement ratio, concrete strength depends on the breaking or the crushing strength of the weakest set of particles in the concrete under the load. These particles would break before the cement bond breaks. Thus, the quality of aggregates is equally important. On a road project, the maximum crushing strength of concrete is attainable in a laboratory tests were 300 kgs per square centimeter due to the constrains in the quality of aggregates. So, bridges and culvert structures were designed as Reinforced cement concrete (RCC) structures instead of pre-stressed concrete structures needing minimum 600 kgs per square centimeter crushing strength.

Cement paste spreads around aggregates and glue them together, hence is dependent on total surface area of aggregates available. If silt which is very fine, having large surface area or too much fine sand are mixed in concrete, the surface area will increase and resulting of weak concrete. Silt is so fine that sometimes cement does not reach all around all the particles, leaving loose pockets in concrete, which is dangerous and makes concrete weak.

Many times aggregates are rejected and ordered to source from long distances at extra time and costs to the project by knowingly or unknowingly not washing and screening aggregates and sand for removal of excessive silt and traces of clay present in them. Aggregates and sand must be washed thoroughly before tests are performed and rejected. This exercise requires some additional equipment and man power at additional costs to the contractor but is always cheaper than sourcing this material from long distances and saving in cement consumption.

8.8.1 RCC

RCC is the most common and safe kind of concrete structures since they are easy to build with least chances of failure. Depending on quality parameters used in construction adequate factor of safety is provided in design of these structures. Basic design consideration is that concrete is good in taking compressive loads and some shear loads. Tensile and shear strength to the structure are provided by re-bars by placing them at locations of tensile and shear stress in the structure. Compressive strength is provided by concrete. Some times if compressive loads are more, than capacity of concrete, re-bars are also provided in compression zones to suport the concrete in these areas.

- Detailed construction drawings are provided by the design engineer with shapes and sizes of all members of the concrete structure.
- Design engineer specifies the strength of the concrete and re-bars that are to be used.
- Accordingly, procurement actions are taken for re-bars and concrete. Rebars must be tested to confirm specifications while procuring.

- Form work or better known as shuttering are the molds made out of steel or timber or combination of both. Concrete is poured in these molds and it takes the shape of structure after becoming soild.
- These molds are assembled by bolts and nuts and removed carefully after concrete has gained enough strength. Normally vertical sides are removed after 8 to 12 hours, bottoms of slabs after five to seven days and beam bottoms after twenty-one days.Decision on Removal of shuttering for slabs and beam bottoms are done by testing concrete samples taken at the time of pouring concrete.
- Scafolding is the suport frame work made of steel or timber or their combination on which concrete form work is fitted and they are strong enough to take loads of concrete and workers moving around for doing the work.
- Formwork and scafoldings are designed carefully for strength and ease of working. In case of repetitive use of form work for over fifteen times, steel from work is fabricated and used.
- Timber forms with ply wood facing are used when the use is limited to seven to ten times. Some times ply wood bolted on steel frame is also used and give results better than steel straight forms..
- Reinforcement bars, cutting and bendings are planned by preparing bar bending schedules from the drawing and properly checked for any mistakes. Efficient bar bending schedules will have minimum wastages.Bar bending schedule is a drawing or statement of how the re-bars are to be bent with all dimensions and sizes.
- Re-bars are cut and bent with proper identification tags are brought to the site on schedule.
- All the above activities have to be done simultaneously by micro planning. these re-bars bent in proper shapes and sizes including epoxy paint if required should be ready before the site is ready to accept them for use and continue to be available in stages according to the schedule.
- First survey is carried out and location of concrete member is marked on the existing structure or ground as the case may be.

Scafolding is fixed as per design at the location, on this scafolding bottom of beam or slab is fixed. One side of the beam is also fixed.

- On this bottom of the beam, re-bars are fixed in proper shapes and sizes as per drawing and bar bending schedule and suported against bending or falling by vertical form work fixed on one side.
- Cover blocks of cement mortar or plastic are provide on all sides.
- Once ready, the forms and re-bars are checked, including shapes and levels. Forms are again cleaned and then closed with proper supports.
- Now, concrete is poured in the forms.
- While pouring concrete it is extensively shaken so that it could flow and reach all the corners of the form work as uniform distribution. and concrete is produced free of any air pockets or honeycoumb. For this equipment called vibrators are used.
- Honeycoumb is a loose pocket in the concrete structure with or without some cement paste, to be removed, treated and repaired. This has to be avoided and excess of honeycoumbs could made the structure liable to be rejected.

Vibrators are Available in Following Types

- Needle vibrators, are 25 mm to 60 mm dia small cylinders about 300 to 600 mm long with an eccentric rod fitted in two end bearings. When this rod rotates the cylinder vibrates. These rods are connected to motor with a flexible cable in a rubber hose pipe about two meters long.when this vibrator is immersed in wet concrete, concrete also shakes and gets spread in in the whole form work and get compacted.
- Form vibrators. These vibrators are bolted from outside to forms. When operated. The whole formwork vibrates and concrete is spread all over in the form and around steel re-bars.

Curing of Concrete

As the concrete surface starts becoming hard, water is sprinkiled to keep it cool and avoid cracking due to heating and formation of water vapors. This is more important when high grades of cement is used since in such cases heat generation is much higher.

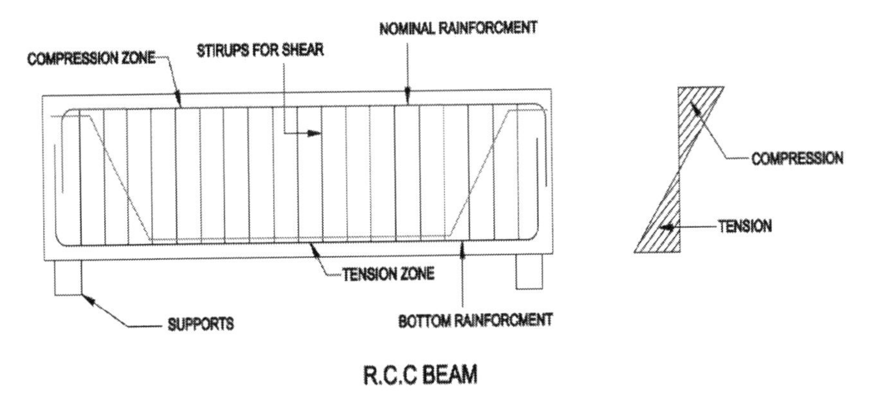

R.C.C BEAM

8.8.2 Pre-stressed Concrete

If we try to bend a green small branch of a plant holding two ends by two hands, it would be noticed that on one side small tissues of the branch come closure and other side they move apart in bending zone. On the side, tissues come closure, since they are compressed called **compression zone** and on the other side they are pulled apart called **tension zone.** In compression zone, length of the branch reduces and increases in tension zone due to pushing and pulling of tissues. Difficult to measure but looking carefully one would feel it. Concrete is a mix of stone aggregates glued together with cement. If one tries to pull the bond of cement would soon break. But if compressed, it would provide resistance and crumble on excessive load. Hence, concrete is **week in tension** load applications due to pulling and **strong in compression** loads applied by pushing between two ends.

Steel rods also have the similar characteristics but they are stronger than concrete. By any process. if one can compress the tension zone of concrete before load is applied. In such a case on application of tension load, first the compressive load would be neutralized by the tension load and would go in tension stress mode **only** after neutralization of compressive loads entirely.

You have a spring. First press the spring (compression mode) and then start pulling (tension mode). You need not to exert pulling energy before spring comes to its original stage. In **pre-stressed concrete technology,** the concrete member is compressed in its tension zone before applying load. On load, induced tension would neutralize the compressive load first till finish of induced compression and then concrete would be required to face tensile loads.

Induced compressive stresses are substantive and even on full loads, structure would still have some compressive stresses in balance in this pre-stressed portion of the structure, hence this concrete of the structure would never be subjected to tensile stresses and property of good compressive strength of concrete is utilized to maximum possibility. Obviously, concrete quality has to be consistent and with very high compressive strengths.

Compressive stress is induced by providing a bunch of special high tensile steel wires in a designed profile before concreting. After concrete is properly cured and gained strength these wires are pulled, to induce compressive stress in surrounding concrete and locked. These compressive strengths are introduced in concrete in following steps for a pre-stressed beam;

- First formwork for the bottom and one side of the beam are fixed in place.
- Pre-stressing cables are prepared according to the specifications and sizes provided in the drawings.
- Re-bars as per the drawing are fixed on the form work. These bars are for extra support to cover contingencies.
- Pre-stressing cables are fixed in the beam in a parabolic shape. Bending moment of a simple supported beam is maximum in centre and reduces to zero in a parabolic order as one moves to end support. Hence these cables are to be placed in designed parabolic profile without any kinks or other mistakes. For convenience of fixing cables their profile is marked on the form work and cables are fixed as per these profiles marked on form work and tied to re-bars.

- These cables are in a thin flexible metal casing called sheeting. Care has to be taken that these casings are watertight and are not damaged in handling.
- Both the ends of each pre-stressing cable is fitted with a female cone. This cone is a circular hollow cylinder of concrete about 250 mm dia and 200 mm long with a corrugated tapered hole in the cylinder. This cone is fitted on the formwork with a matching hole to the hole in cone. These cables are passed out through this hole. At least one meter of cable wires should be out of forms on both sides of the cable.
- After checking that all the cables and re-bars are properly placed, the formwork is closed and made ready for concreting.
- Concrete is poured in the forms carefully and make sure that it reaches all the corners of the formwork including cables. For proper distribution of concrete and its compaction form vibrators are preferred. Needle vibrators may damage the sheeting of cables and being a thin structure may not reach everywhere.
- In spite of taking all the care, the sheeting of cables could have damaged somewhere and cement grout entered in cables and create a bond. After a few hours of concreting cables are hammered gently to break any obstruction in the cable.
- Concrete would need about ten days to achieve minimum strength required for pre-stressing. Start pre-stressing only after sufficient strength is reached, confirmed by laboratory tests.
- Clean the pre-stressing wires properly and again check by hammering that they are kept loose in the beam's cables and insert male cone loosely fitting in female cones, keeping wires evenly distributed in groves provided between the two cones.
- Fix pre-stressing jacks to the pre-stressing wires on both sides of the cables.
- Pull the cables from both ends by the jacks and provide required tension and check elongation of wires. Elongation is checked by putting a mark on one or two wires before start of pre-stressing and

check distance of this mark before and after tensioning from face of concrete.

- After pre-tensioning is completed, push the male cone hard in the female cone with a piston in the pre-stressing jack and make it tight fitted.
- Release the jacks and repeat the process for other cables as instructed.
- After a few days check for any slippages in wires, if there is any slippage, break the male cone and repeat the pre-stressing of that cable.
- It would be noticed that the beam lifts by a few centimeters in its centre; this upward deflection is measured and checked with expectations as per drawing.
- Grout the cables with a non-shrink grout for pre-stressing wires to be integral part of the beam and protected from corrosion.

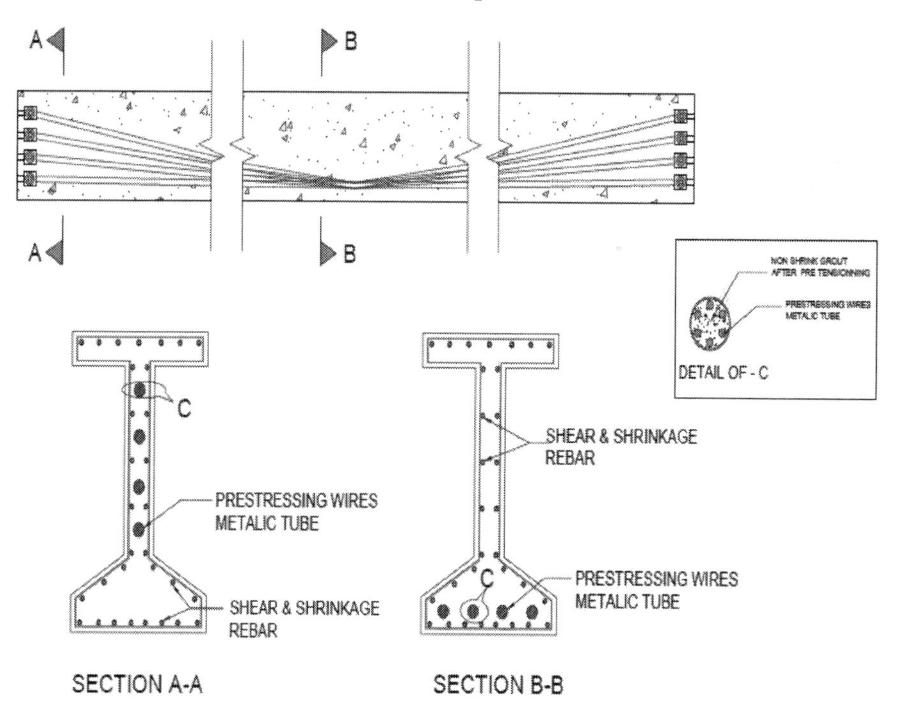

With this, pre-stressing is completed. This is as per Freyssinet system of pre-stressing of concrete. Many more systems are available at the choice of designers but concepts are the same.

In some cases, pre-stressing is done in two stages. In first stage enough compression load in concrete is provided in beams to take care of their own weight and handling loads up till they are placed in their final location. In the process, tensile loads created due to weight of beam, called dead load, nullify majority of pre-stressed compressive strength. Now a few more cables are pre-stressed to take care of live loads. This is marvelous but concrete has to be very good quality to take multiple stress reversals. Pre-stressed concrete in beams and other structure is preferred to RCC since it is almost half the weight, easier to handle and cheaper in construction.

8.8.3 Roller-compacted Concrete

Major dam projects have a wide base that slowly tapers and reduces in width as it goes up in the designed shape of a huge mass. These masses are built to provide a big thick wall, stable with its gravity to resist the pressure of water on one side of the wall and the other side remaining free. These dams are built by Earth fill, rock fill or concrete. In most cases, Earth as well as rock is locally available. Some rock also comes out of the excavation and is used for the project in some form or the other.

In normal concrete for structures like concrete dams' cement consumption is more than ten percent of total concrete weight with 75mm being maximum size of aggregates and elaborate infrastructure and construction procedures are needed. With the recently developed roller-compacted concrete technology, graded aggregates in 150mm or 100 mm down in sizes are mixed with 3 to 5% cement and added with very little water as per designed water cement ratio. The mix is almost dry but no voids should be there after compaction as confirmed by laboratory tests. This aggregate, cement and water mix is spread by trucks and motor graders in layers of 300 to 400mm across the dam section and rolled by heavy rollers for a few minute and completed before initial set of cement concrete. This is very fast as compared to regular concrete and Earth work construction. Cement content is less than half or one third of regular concrete of similar properties and working procedure is less than one third time taking compared to Earth work. Section of dam is also very small compared to earthen dam. This technology is evolving fast and will be used in the construction of very heavy and mass concrete structures. Since the cement content is less, problems of handling excess heat generation by cement are reduced and its bad effects on concrete are eliminated in roller-compacted concrete.

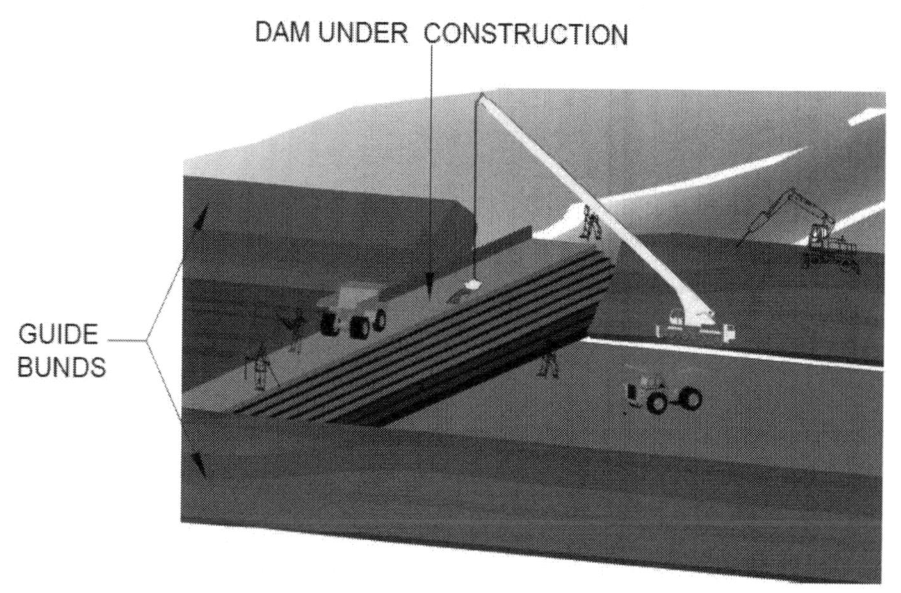

Roller-compacted concreting on a dam site in progress

8.8.4 Plum Concrete

In some mass concrete foundations, boulders of say 150 mm size up to 30% of total quantity of concrete, are mixed in the wet concrete. This adds to economics and save heat generation. Such concretes are called plum concretes. After at least two layers of wet concrete is completed, well-shaped and thoroughly washed, angular rock pieces with a maximum size of not more than 60% of layer of pour thickness are placed in the concrete, keeping a distance of more than one-and-a-half times of the biggest size of aggregates in the concrete. The bottom of the boulders should be embedded in the wet concrete to avoid voids at the bottom. Concreting is continued without stopping. These rock pieces become an integral part of concrete structure.

8.8.5 Colcrete - Underwater Concreting for Large Foundations

Colcrete is a method used for concreting a thick block in a confined area in deep underwater conditions. Colcrete is a process of concreting by injecting with high pressure cement water or cement water-sand grout in voids of a big container filled with loose aggregates or stones under deep water. This grout is fed in highly agitated condition and **does not allow chemical action between cement and water till agitation is reduced or stopped**. This agitated state of any fine solids mixed with water is called colloidal state. Since concrete is done in colloidal state it is called **colcrete**. It is mainly used in concreting plug of deep well foundations and foundations of marine structures.

Iron ore in powder form is transported hundreds of kilometers in colloidal state with water through pipelines in Brazil and India. If colloidal condition breaks, solids would immediately settle down and pipe running in more than 300 kilometers would be choked.

In colloidal concrete process, the whole space to be concreted is first cleaned and enclosed. If required help of divers is taken. These divers would go down in deep water with help of compressed air, through a pipe for breathing and fix properly the form work of the foundation. Once the form work is ready, it is filled with aggregates. While filling with aggregates, a few 75 to 100 mm dia pipes are left loosely embedded in the aggregates at 3 to 4-meter spacing, keeping the bottom of the pipe at the bottom of foundation and the top above water. Grout in colloidal form is injected at a pressure of at least 3 bar plus head of water in the area to be concreted. This grout is either neat cement grout mix of water and cement with water/cement ratio as low as 0.35 or sand cement mix grout with sand/cement ratio of 3:1.

There are special colcrete grout mixer machines. They have two mixing and agitating drums. In the first mixer, first, water is filled and agitated. Then, cement and sand are added while water is still agitating. Once water, cement and sand are uniformly mixed, the whole mix is transferred to the second drum agitating with much higher speed and pumped into the aggregate base through one of the pipes. The whole process takes less than five minutes. The grout spreads in the base first and slowly comes up replacing water. The grout is pumped through all the pipes in rotation for uniform distribution in area to be concreted. For large foundations, multiple pumps are operated simultaneously. In this process the grout is pushed from bottom to top hence water does not wash the grout except some at top which is ultimately washed away. Colcrete grout mixers are generally patented and available all over the world.

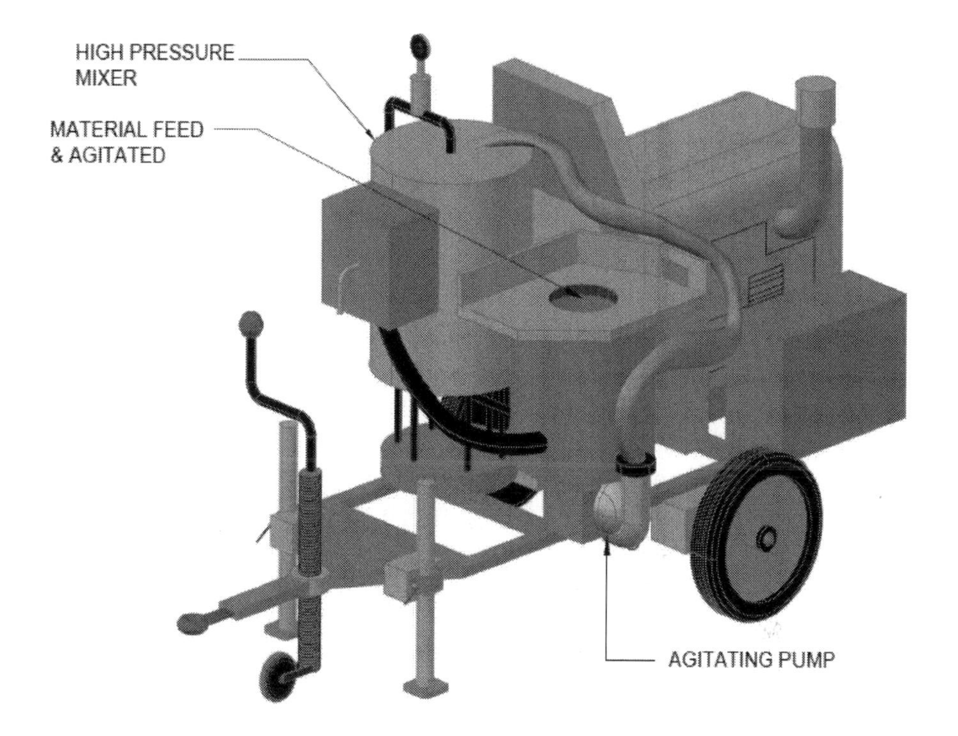

Pipes are slowly lifted as required and removed after pumping operation is over. Within two to three days, the concrete is set.

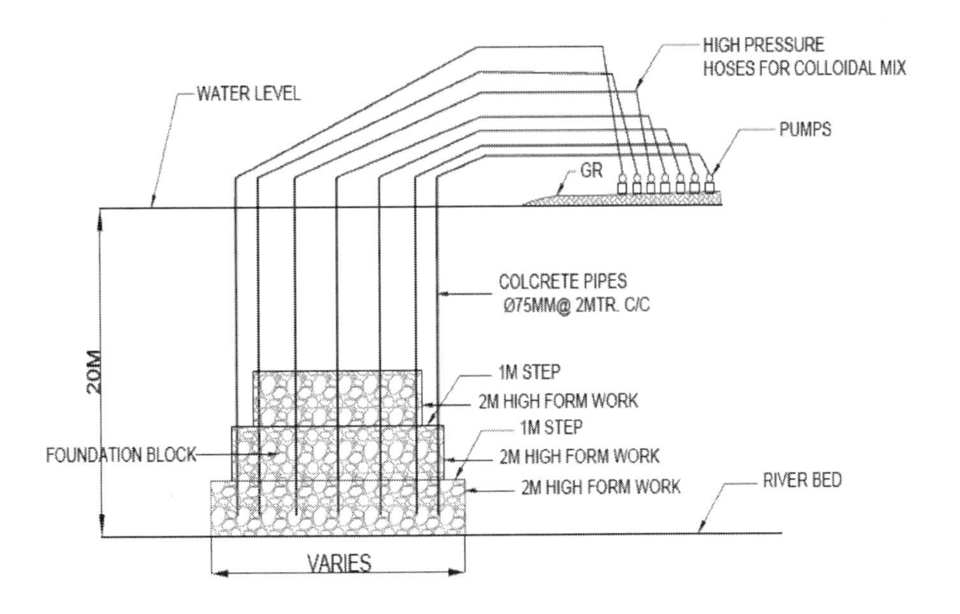

A large subsea foundation to be filled with colcrete layout is shown above. Multiple pumps are used due to a large foundation and each pump will feed a full row of the foundation with a number of pipes as required. It would be preferred to plan such colcretes in stages of two meters' height, unless unavoidable, since fixing formwork, re-bars and other preparatory works are easier to manage with maximum of two meters' height under water.

For higher foundations, a complete prefabricated form has to be lowered in the sea by a crane. This form should have enough stiffeners. First, while lowering the form in water, there is heavy turbulence which shakes the form. Secondly, when filled with rock, the rock will hit the supports and likely to break them. Thirdly, if the ground is not prepared, the level and form would sit unevenly and extra stress would be developed. These factors are to be considered in fabrication and handling the form work. On one of the port projects, a foundation with the pile was to be built under water for concreting raft foundation over piles, a big steel formwork was made. The site engineer designed and fabricated the form with steel plates and timber struts jam fixed inside. While lowering the form, due to turbulence, all the struts became loose, and collapsed within the form. The form was broken badly and rejected.

8.8.6 Tremie Concrete-Under Water Concreting for Small Foundations and in situ Piles

Tremie concrete is a procedure to carry out concreting operations underwater for small deep foundations, columns and cast *in situ* deep piles on all locations containing a high portion of subsoil water in the piles and similar structures in the prefixed forms. One has to take care of following while doing underwater concreting:

- Concrete should not be segregated. This is achieved by conveying the concrete through a narrow pipe and entire mass travel together and deposited in its confined place.
- Concrete as finally poured, should not mix with clay and other rubbish at the bottom of the foundation and not washed by the water around.

To take care of both the requirements, Tremie concrete is done for foundations as explained hereunder:

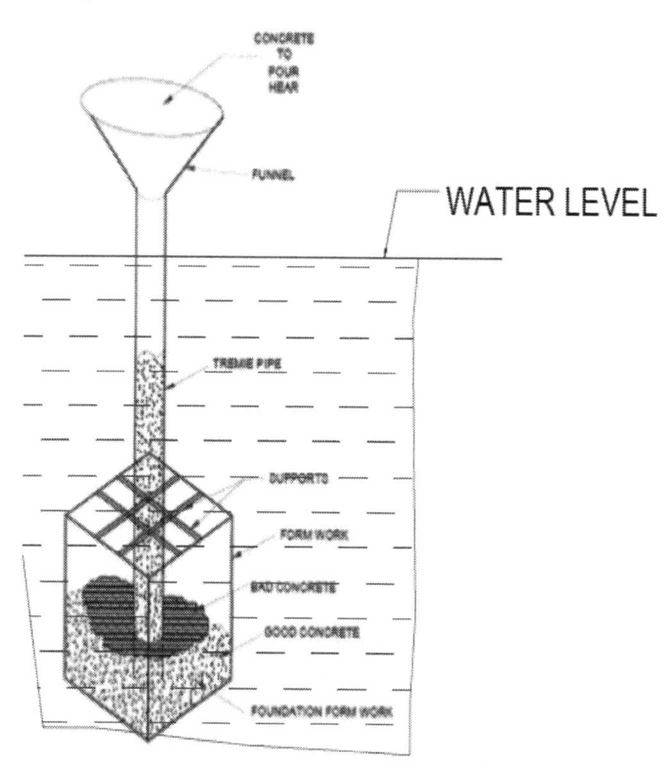

- In this system, to concrete underwater, a steel pipe is fabricated with a funnel on top as shown in the sketch. This pipe is called Tremie pipe. The total length of the Tremie pipe is the depth of the foundation or pile, plus a few meters to always be above the ground or water. For deep pile foundations, Tremie pipe is made in sections of about three meters and screwed together. The diameter of the pipe and the size of the funnel are decided on the basis of the quantity of concrete when it is full in the pipe. The funnel should be equal to at least half meter in height of the concrete in the foundation. But the diameter of the pipe should not be less than 125 mm.
- Once the foundation is ready for concreting, the first rebar cage is lowered in the foundation.
- The Tremie pipe is lowered, touching the bed of the foundation. Water in the pipe will be the same as outside.
- The funnel of Tremie pipe is connected to the hook of piling rig or a crane as the situation may be.
- A rubber ball fitting tight in the pipe is placed in Tremie at the neck of funnel and tied with a string long enough. Start pouring the concrete in the funnel, holding the string firmly.
- Release the string, a bit - say one meter - at a time. The ball will move down, holding the concrete on itself. Water will not enter the concrete in the pipe since the ball is a barrier between the concrete and water.
- Continue the operation till the ball with concrete reaches bottom of the pipe and the funnel is also full with concrete. Release the string holding ball and slightly shake the pipe. Due to the weight of the concrete ball, will escape and come up afloat in the water.
- As the ball is released, concrete in the pipe will also be released and will be spread in the foundation, leaving at least thirty centimeters height of concrete around in the pipe, same as outside the pipe. This concrete will sweep the floor of the foundation and push above with all the rubbish mixed with first layer of the concrete.
- This will be followed by good concrete through the pipe as concreting proceeds.

- Concrete is passed through the Tremie pipe since the end of the pipe is on the ground. The concrete will come out of the pipe at the bottom of the foundation and move upwards, pushing the earlier concrete and bad concrete mixed with impurities, will always be on top. After some time, concrete will not go down. Lift the pipe a little and the concrete will slip down. This process is continued till the end. Bad concrete is either washed away or is found in the upper portion of the pile. It should be broken and discarded.

Tremie concrete is not advisable for large open subsea foundations since it would be difficult to always maintain the column of concrete in the Tremie pipe, resulting in bad concrete mixing with good concrete. Thus, colcrete is preferred for large foundations and Tremie concrete for small foundations and *in situ* pile concreting. Still, if colcrete system is unavailable, a big foundation should be divided into manageable compartments and concreted one after another, by the Tremie method.

8.8.7 Slipform Concrete for Vertical Structures

As its name, slipform concrete is a method of concreting, form work, the formwork of concreting keeps on slipping upwards on sides of freshly laid concrete as **concrete is poured for days together without stopping or changing the form work.** In the process enough wet concrete is maintained in the forms all the time till required height of structure is achieved and concrete pouring is stopped.

If concrete is to continue after a break, make it ready for **next concrete** now itself. After stopping concrete, wet concrete is leveled in the forms and extra if any is discarded. Slipforming still continues for about 90 minutes till all the concrete is set and released from forms. Simultaneously, all the forms are leveled accurately and cleaned with proper tapers in all the walls. This should not be left for the next day and necessarily a fresh set of people have to be organized in advance for this purpose to step in immediately after stoppage of concrete. For a tapered chimney structure or cooling tower, many patented designs of form works are available. Here we will discuss only vertical structures since the principles of slip forming are the same for all type of structures.

In slip form concrete, a number of hydraulic jacks each having capacity of say three tons are fixed to the special forms. Total load required to be lifted while concreting including friction loads and men working on top is equally distributed to sufficient number of jacks fitted and well distributed in entire form work.

Slipform Jacks

Slipform jacks are hydraulic jacks. They climb on a 30mm steel rod in strokes of 50mm at fixed intervals varying between two minutes and five minutes, depending upon speed of concrete supply and initial setting time of concrete. if concrete supply is slow, form work is lifted slowly and slipforming cycle of lifting formwork is changed. Similarly, if concrete is not setting fast, forms are lifted slowly. These jacks have two sets of jaws, one at the bottom and another on top of jack. There is a vertical hole in centre of jack for a heavy duty pipe about 30 mm dia termed as slipform rods is fitted in the jack through this hole. The jack climbs on this rod only in upward direction holding the rod firmly by one of its jaws. These slipform rods stand vertical on base of the structure in four to six meters' heights and are extended as concrete progress by joining new pipe with threaded studs.

When hydraulic pressure is given to the jack, cylinder between the two jaws expands by 50 mm by pushing down the piston and bottom jaws. And bottom jaw moves down by 50mm, keeping tight grip on rod and pushes the jack upwards. When full pressure is applied to all jacks, hydraulic pump trips automatically. On release of pressure, bottom jaw would move up by 50mm to its original position. At this time upper jaw would hold the climbing rod and not allow to slip. Pump would again start as per timer and again jack would climb by 50mm. Jack keeps on climbing on the rod step-by-step. This exercise continues till concreting is finished. This process is similar to a man climbing on a coconut or palm tree, embracing the tree firmly with his hands and feet. Lifting of form less than 50 mm at one time is not desirable since every time of lift formwork has to break any bond with wet concrete for this minimum lift of 50 mm at one time is desirable. However, in some brands of jacks this could be adjustable.

Form Work

Concrete takes about forty-five minutes to set or become solid from time of pouring in formwork. Forms are one meter-high and 750 mm concrete is always in the formwork while concreting is in progress and is continuously poured in the form. Considering 50 mm lift at every three minutes. For moving concrete out of formwork from the time of pouring, it will need fifteen strokes every three minutes, so that makes forty-five minutes. It means concrete is set in the formwork before it comes out of it. For speed more than three minutes, concrete speed has to be fast and all the forms should be always almost full up to almost one meter instead of 750 mm. Reinforcement steel bars are also tied simultaneously as concrete progresses and forms move upward. These bars provide additional support to concrete. The speed of lifting the forms is adjusted according to atmospheric temperature and supply of concrete as stated above.

Concrete once comes out of form should not bulge as well as should not be slow and stick to the form. Form must keep on moving without stop and slipping on concrete. If speed of lifting is slow and concrete sticks to form, the friction created between concrete and form is much higher than tear strength of concrete at that time; hence the jack will still lift the form and tear off the concrete, carrying set concrete with it. If not repaired immediately by releasing concrete from form by breaking with crow bars, jack would continue to lift set concrete with increase in quantity every time, increasing load on the jack and ultimately jack stops working due to excessive load. This would spread like cancer and within a few minutes the slip forming operation would collapse. For cold climate speed of the form work to be reduced for allowing initial set of concrete within the form or some additives are added in concrete.

A Few Important Issues for Slip Form Concrete

- Consistent supply of quality concrete for long duration in a single operation.
- Adequate supply of raw materials to batching plant on non-stop basis from stocks, hence full quantity of all the material must be in stock before start of concrete and available till finish.

- Re-bars cut to size available at site with arrangement for continuous fixing as concrete proceeds.
- Change of shifts and relievers on job for work force including management. There was a failure due to an overworked expert going to sleep for two hours.
- Distribution of concrete uniformly in all the forms.
- Power backup and repairs management.
- Ensure all the jacks are going up with a uniform level of forms. In case of any distortions, should be rectified soonest possible.
- Continuous trowel finish and touchups to set concrete outside surface.
- In case of problems, planned shutdown to be taken under expert directions and supervision.

If done properly, seventeen-meter-high sea port foundation having twenty-eight vertical compartments with almost 220 jacks, working together could be finished in less than two days with consistent supply of concrete average twenty cubic meters per hour and re-bars twelve tons per hour in floating condition including time spent for startup and closing down activities, way back in 1972 in Visakhapatnam Port, for construction of ore berth and then this technology continued in India on many port projects for many years successfully. For taller concrete structures, planned breaks are necessary due to limitations of structural strength in fresh concrete and management constrains.

Form Work Assembly for Slip Forming

We have shown typical formwork for a structure having walls of 250 mm thickness. Form work is preferably made with close-grained seasoned timber of 40 mm × 125 to 150 mm × 1 mtr long planks, free of knots. These planks are joined together by tongue and grove joints nailed on two hard wood seasoned and straight runners 150 mm × 100 mm in section. In case of curvature in the structure, like circular sections or corners of a caisson, two numbers of 200 mm × 50 mm. Timber planks are cut to the desired profile and nailed together with staggered joints. All timber planks of 40 mm thick must be nailed in the form keeping timber vanes going from top to bottom to avoid extra friction by timber on fresh concrete and its fast wearing out. If one moves his hand on the plank upside down, it would be along the vanes and smooth feel but if he moves in opposite direction ends of vanes would create friction and hurt. All planks and 150 × 100 timber runners should be planed straight on both side with uniform thickness is maintained. Keep a clear distance of 200mm from the top of the form to the upper runner and the same from the bottom of the form to the lower runner.

Take the forms on the raft foundation and assemble on the level surface in position with temporary supports. If the wall thickness is 250 mm, the spacing between the forms in the bottom would be 252 and 247 mm on top. Forms are fixed in their conical shape to reduce the friction while concrete is moving in the form. Fix the yoke and channels as per the design of the slip form equipment supplier and nail them properly to the lower runner of the forms. Jack will be loosely but firmly placed under the bottom of channel of yoke assembly. While working, jack will push the channels upward. In turn, the channel will push the yoke assembly to move upward along with forms and entire infrastructure leaving concrete in place. On top of form work there would a deck for working and other fittings would be generally as per the following drawing, with small changes as per requirements of the suppliers of equipment. Covered deck in addition to work platform, arrests the heat generated by fresh concrete below the form work and helps in fast gaining strength after initial set. Timber forms are preferred to steel forms due to following reasons;

- Steel forms become very hot and difficult to work with after some time
- Linear expansion of steel due to heat is more compared to timber
- Timber forms are easier to repair while concreting is in progress
- Steel forms unless made of thick plates, which are heavy, gets kinks in handling, not acceptable for slip forming, since fresh concrete would get into kinks and provide extra friction.

If handled carefully, steel forms could have multiple uses, while timber forms are good for one and second use if handled very carefully and not damaged while slip forming.

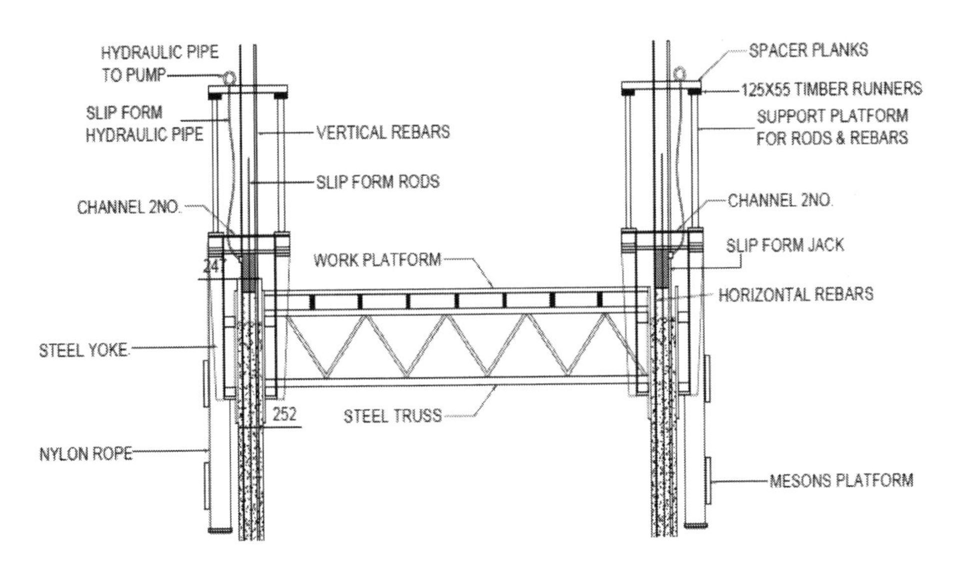

Troubleshooting and Precautions in Slip Form Concrete

Before we start out with pouring concrete, the following things are to be checked:

- That all materials required for the pour are available on site with extra backup.
- That all equipment is properly serviced with fuel and lubricants available sufficient for the operation.
- That all labor and supervision staff are properly planned in shifts with definite logistic arrangements.
- That proper lights for the night are organized. Do a trial the previous night.
- That concrete equipment like buckets are serviced and in good shape.
- The level of the form by leveling the instrument or water tubes.
- The taper of each form at least at two meters' distance and correct if any discrepancy.
- The hydraulic tank of slip form system with all the pipes full and bleed it for any air pockets. For air bleeding pump is started at very low pressure without disturbing the forms.

- All the slip form rods are standing firmly on base and jacks are touching the channels of yoke assembly. Rods go down with little push by hand.
- That necessary steel re-bars are fixed.
- Wet the forms with a good water spray and clean if any rubbish around.
- Keep a sufficient number of crowbars for emergency on the deck, distributed at a few locations.

Now start the concrete in 150 mm layers spread uniformly in the entire form, First, liftoff 50 mm to the form is given after twenty minutes or half hour maximum of starting concrete. Attend to any discrepancy and proceed and continue as stated above till finish of work. There may be some reverse slopes. in such case concrete would crack. Repair formwork immediately.

Concreting of a float precast concrete foundation (Caisson) is in progress in Goa, port in India in 1974.Concrete is poured by a bucket fixed to a floating crane on right side. On the left is barge carrying steel bars. Vertical lines are for re-bars and slip form jack rods. Top horizontal line is supports for re-bars and rods, keeping them straight. Then two horizontal lines show the slip form. Bottom line is for masons to inspect. A hut on the right of the concrete bucket is the hydraulic pumping station to which power is fed by generators on crane barge

Two concrete caissons sit on a prepared sea bed in 20 meters Deepwater and further work on top of these caissons is in progress.

8.8.8 Slip Form Concrete in Road Pavements

Roads and highways are classified in two categories. Rigid pavements such as concrete roads, and flexible pavements such as asphalt roads and others like Water Bond Macadam roads and such else. Concrete roads are also called rigid pavements. They are durable if the sub-structure up to the base course does not settle or sink for any reason including landslides, poor construction, inadequate drainage etc. In case of settlements, the road cracks and a big part of the road's concrete is cracked and settled, hence discarded, removed and redone. It needs to be redone after the removal of the affected concrete with heavy equipment. In case of flexible pavements, the top of the road adjusts with the settlements and the road is still usable, but it will not be comfortable, since it will be an uneven surface. Repairs are cheap. It just needs the filling up of depressions with asphalt concrete or approved material. For new roads, unless very good quality control measures are adopted, it is advisable to make flexible pavements and after a few years of good usage and settlements, if any, to convert to rigid pavements. For concrete roads, the following things must be done:

- The road foundation up to the base course is prepared. This is common for flexible and rigid pavements.
- Concrete pavements are made in *in situ* concrete slabs of a required thickness, touching each other with the provision of expansion joints as designed.
- The top surface of these concrete slabs is finished manually and despite all the care, some level differences are unavoidable.
- These uneven levels in the road surface are not comfortable for driving and are noticeable because of small jerks.
- A few engineers provide a 50mm topping of asphalt for better driving comfort. This takes care of concrete uneven surfaces and wearing off of weak road surface due to weak concrete due to segregation while pouring, bad curing and inadequate cement content.
- Due to direct dumping by transit mixers on ground, big stones get separated from the concrete and are thrown a little away. Only slurry drops in a heap on the ground. The masons will drag these stones in the concrete but the mix is not homogeneous. This is called segregation of concrete.
- Segregation of concrete reduces the quality and some weak pockets are formed.
- These weak pockets create potholes on the road after sometime. It causes a question mark on the capabilities of the contractor, due to negligence of the mason and the supervising engineer. It takes only a few minutes to mix the wet concrete by a few people with spades.
- Now, slipform concrete pavers are available. Concrete is poured in the hopper of the paver machine from the back and the paver spreads concrete evenly in the front, while moving back for the full width of the road.

- The level of concrete surface is monitored by a fine metallic wire attached to a slipform paver, which maintains a uniform level of concrete as per the profile of the road. a sensor travels on wire as paver is moved and adjust concrete paving height, keeping top of concrete in level of the wire. Better is the wire profiles in firm supports, better is concrete surface.
- The concrete is free of segregations since it is mixed again in the paver machine.
- Here supply rate of concrete has to be large and should be maintained throughout the paving operation.
- This system gives the road a machine finish. It is smooth, leveled and durable. The entire section of the road concreting can be done in one operation and can complete up to half kilometer of the road in one shift.
- Slipform speed is to be maintained at a minimum of thirty to forty meters per hour for the whole two-lane carriage way. In case of constrains in concrete supply, half the road width is taken at a time.
- Concrete roads last longer than eighty years without maintenance if done properly. Though 25 to 50 percent more expensive, it is overall a cheaper and a preferred option if investment is available.

8.9 Dredging and Underwater Rock Blasting

According to the logistic and other considerations of turbulence and waves in sea if favorable, new ports are built in the deep sea, and are connected by a long Approach Bridge to the land. However, to cater for turbulence and waves for a new port, generally breakwaters are to be built which are very expensive.

Alternatively, ports are also built on land a bit away from sea. A water channel for bringing ships to port and a deep lagoon are created for movement of ships in the port. This work is done by digging and throwing away excavated Earth or filling in low land around the port. Generally, where the sea is very rough and not safe for berthing of ships and suitable land is available, this approach is adopted. Water remains protected by land all around hence, there would be no effect of turbulences and waves in the sea on water near the port and ships berthing alongside is comfortable.

For construction of new port, Earth is excavated and carted away by making a deep water channel for movement of ships from deep sea to port, Simultaneously, Earth above the water level is excavated with normal methods and is carted away by normal trucks to low level reclamation area. Then, a few excavators with a long reach under the pads are used to remove maximum Earth, starting from the seaward side and then move backwards, making a few meters' deep water channel. After this, a dredger of an adequate size as available will move in and excavate the Earth underwater up to the required depth. Underwater excavation is called dredging. These dredgers are of two types: trailer suction dredgers and cutter suction dredgers. They are big ships, made for dredging purpose. Interesting to bear in mind excavation by backhoes and trucks is done only till the dredger is arrived and then it will cut the whole thing very fast.

Trailer Suction Dredger

Trailer suction dredgers suck the Earth, mixed with water, while moving the dredger in a water channel. A suction pipe slides on the bed of soil, taking the soil and water to its holds, and then dredger moves to the dumping area in the deep sea, about a few kilometers away. Dredged material is discharged and the dredger returns for the next load. Each load of over 100,000 tons with over 30% soil content takes ten minutes to load.

Cutter Suction Dredger and Reclamation

In case of hard soil to be dredged, or dredged soil has to be used for reclamation a cutter suction dredger is used. This dredger has a circular cutter near the suction pipe and cuts the soil with its cutter, while it is sucked by the suction pipe. Cutter suction dredger has a provision to discharge the dredged material through a long pipe to the dumping area or in a series of barges, to take the dredged material to dumping grounds. These arrangements are done on the basis of the availability of equipment and logistic considerations. Disposal by barges is avoided since it is slow cumbersome and very expensive. In cases the dredged material is to be used for reclamation, the discharge pipe is connected to the reclamation area where dredged sand is filled mixed with water and naturally get compacted. Sand attains maximum density in saturated condition. The discharge end of pipe is shifted to different locations at the time dredging stops for maintenance. In cases reclamation area is too far, booster pumps are installed at required locations in the discharge pipe.

Rock Dredging

Some hard patches, in sea bed of entrance channel or in the port, considered not possible to be removed by cutter suction dredger may be left on the bed and have to be removed for the free movement of ships. These patches are drilled under the sea, loaded with explosives blasted and dredged away.

This Procedure Called Rock Dredging and Works as Under

Decide the depth of water in high tide to which blasting and dredging has to be carried out. Select a big barge about 600 tons' capacity with good moorings. Moorings are the ropes with anchors connected to four or six air winches fitted on the barge to keep the barge stable in water. Anchors are thrown at distances in different directions and barge moves a little, say twenty to thirty meters in all the directions and held stable by tightening or loosening ropes to these anchors through the respective winches. These ropes are loosened for movement of barge and on positioning to next position they are again tightened, but never left loose. Take a drill tower generally fabricated with aluminum box sections and mounted on two rails fixed on one long side of the barge. These rails are usually two meters' apart tower will have four wheels to freely move on these rails from one end to other and held firmly in position by steel ropes about 12mm dia, four numbers. Height of tower would be at least sum total of depth of water after dredging in high tide plus two meters for over drilling plus free board of barge plus space for drifter. Drifter is the drilling machine, mounted on top of drill rod and hung from top of tower with a rope. Drifter is equipment like drill machine works with compressed air. It rotates and hammers on drill rod. Further there is a through hole from drifter, drill rod and drill bit, through which air is passed, up to bottom of bit to keep the hole clean. Typical drill tower is shown in the photograph below, mounted on barge. Drill tower has a guide on sea side for free movement of drifter and two air winches mounted on base of the tower to control drifter and casing pipe movements. Drilling is done through a casing hung by the tower and operated by second air winch. Following equipment are further mounted on the barge Compressor of adequate size, say 600cfm capacity A generator for lights, say 40 kWh A cranes minimum 90 tons' capacity with orange peel grabs.

The Support Equipment Includes:

- Tug boat
- Barges to carry blasted material to dumping area
- Exploder with certified blaster
- Total station with a well experienced surveyor

The System Works as Follows

- A base line is established on the shore line for the positioning of the barge. The entire area to be blasted is plotted on a grid scale. Surveyor with a total station on the base line and barge foreman on the barge will position the barge at the place of drilling and blasting with the help of tug and moorings winches on barge.
- Position the tower on the first drill location. Lower the casing pipe through the drill tower. With a few blows of free fall, let it go through loose silt and clay and stand on hard surface in the seabed, keeping the top of the pipe above the barge level.
- Now, lower the drill connected with the drill bit at the bottom and the drifter on top through the casing pipe, once it reaches the ground, start the drifter that drills the hole in the rock up to the required depth, plus two meters. While lowering the drill rod in the casing, maintain the flow of air through the rod and drill bit to maintain the cleanliness in the drill hole and around the drill bit. Extra two meters is kept since the rock breaks in V shape at bottom of the hole and keeping two meters extra we get a level surface of blasted material at desired level.
- Rock is often drilled by a 75 mm dia bit, and two meters spacing after drilling is completed. The drill hole is air flush cleaned. Take the drill bit and rod out of the casing.
- Charge the hole with a detonator and marine explosives with the help of compressed air up to top of rock with a charging kit with flexible pipe. It goes right to the bottom of the hole. A Hold the detonator wires firmly and safely remove the casing, slowly. Casing pipe has holes at intervals to help retrieve the detonator wire. Move to the next location and repeat for the full length of the barge, keeping the spacing of two meters between the holes.
- Stemming with clay or other similar material is not done in deep underwater blasting since water column of more than 15 meters provide a minimum load of 15tons per square meter on top of rock surface would necessitate explosives filling up to top of hole, the rock surface for effective blasting. Further water column would also provide the necessary cover to explosives.

- After all the holes are drilled and loaded, connect the detonator's wires in a series and end the series with a blasting cable. Tie it to a float to keep all joints above water. This float identifies the location of the blast. Move the barge away to safe distance.
- Blast with the help of an exploder. It will produce a loud sound. Water will rise like a fountain in the blasted area.
- Move the barge back to the location and dredge the blasted rock with the crane fitted with an orange peel grab, load and cart away by barges provided for the purpose. Orange peel grab is a special grab for rock dredging. It has four fingers to grab stone pieces, hits the surface with four sharp teeth and bites into the blasted rock. When closed, it looks like an orange. Each of a four fingers is shaped as one fourth the peel of an orange. This procedure is repeated till the end.

8.10 Pile Foundations

For any structure to be built anywhere need a foundation to stand firmly. If ground is not strong enough to take the load or water body of river and sea need something strong to support the structure. Pile foundations are very important and used widely for building foundations for various types of heavy structures in all the situations. They have made a big contribution to civil engineering design and construction for a long time.

8.10.1 Timber Piles

Timber piles are very handy and helps a civil engineer in a vast range of applications. Timber bellies are the trunk or thick straight branches of trees with a circumference of about 300 to 600mm at two feet from bottom. These bellies are cleaned, branches are removed and taken as straight pieces without any projections. Timber piles are driven keeping thinner portion down for better load bearing results. Otherwise hole in ground would be big and pile dia would keep reducing with a result, pile would be loose and a waste. These piles are load bearing and friction piles as per the ground strata. Piles are driven into the ground with a drop hammer, supported by a tripod till they refuse to go down. They give bearing capacity varying from three to ten tons each. Timber piles are used for ancillary works like minor bridges, small jetties, scaffoldings, buildings and roads. These structures could be built with the least resources in or near forests. Majority of major projects are in logistically difficult areas. And there are challenges to reach either by surface or water.

If a project is near a river or the sea, a small jetty can be made with timber for unloading the equipment and materials. Just drive a few piles on the shore and build up the structure with timber, using horizontal and cross bracings with timber, at about two meters' intervals. Make sure that your jetty does not move laterally due to hitting by a barge. For this, a number of pile rows should be more and cross bracings should be embedded in the ground as much as possible - to the extent of one-and-a-half meters' maximum. Similarly, while making a temporary road, there would be a few big drains or river lets. Cross them with timber bridges on timber piles made in a few days' time. In saturated and slushy grounds, warehouses and offices could be made elevated from the ground resting on a few piles driven in ground by a tripod and hammer well connected by horizontal bracings. Drawing shown below is a typical unloading jetty for equipment and materials from a barge.

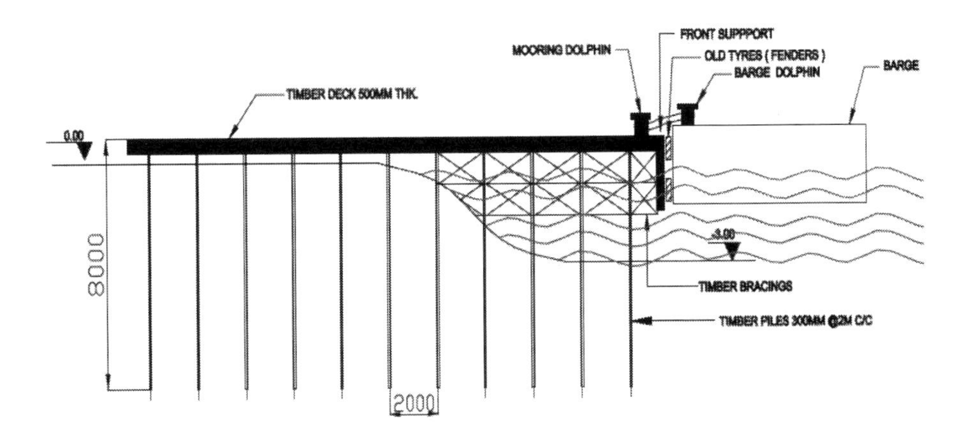

In Indonesia and other countries, short piles are used as permanent ground improvement for major haul roads. These piles push the soil around and compact while pile driving. In addition, load bearing capacity of these piles is also utilized for road construction by driving piles in a grid and join together by horizontal members of timber. The whole area is then filled with sub-base material.

8.10.2 Steel Piles

Steel piles are available in two types

Circular piles

Sheet piles

Circular piles are heavy duty pipes of large diameters with thick wall. these piles are driven by a vibro hammer in deep sea foundations. Construction of sea port structures with steel piles is very fast. One can build a port in 25% of the time with steel piles and structure, compared to regular concrete structure

Steel sheet piles are mainly used for two purposes. They are used for retaining wall for different applications to support the Earth, like port jetties, deep excavations for industrial structures, temporary cofferdams for bridges and river training works. These piles, also called sheet piles, are in different shapes like, Z, C, and others as per project requirement. These Piles interlock with each other and provide a continuous thick steel wall, deeply embedded in the ground to act as a retaining wall. These piles are driven into the ground by vibratory or impact hammer or a combination of both. These hammers weigh about three to ten tons and are operated on a compressed air or hydraulic oil under heavy pressure. Piles are driven in a row, one after another keeping interlock intact. Circular piles are used for foundations mainly in the deep river or a sea, driven by impact or vibratory hammers with a crane on a big barge, for various applications, like oil platforms, sea ports construction and bridge foundations.

8.10.3 Precast Concrete Piles

Precast concrete piles are *in situ* driven piles. These are used in small works. These piles are cast in advance in standard lengths and difficult to extend. Secondly, a casting yard is needed with a crane. Extra reinforcements are required for handling and concrete should be at least M40 grade. For lifting and handling, these have to be handled by a crane lifting at two locations, L/4 from ends of the pile, while L is total length of the pile. While driving the pile, often, the head of the pile breaks, becoming matter of concern and reduction in depth of pile. It is thus, overall more expensive when compared to other concrete piles. Only main benefit is that heavy pile driving equipment, cumbersome to mobilize is not required. And work can be managed by a crane and equipment mobilized for other purposes. These piles carry loads up to 80 tons easily.

8.10.4 Driven Cast in situ Piles

These are friction piles. A standard length of pipe is driven by a drop hammer into the ground by a piling rig with the bottom closed. Once it reaches the required depth below the ground level, the reinforcement cage is lowered in the pipe and is concreted. While concrete is poured into the pipe, the pipe is lifted with a few strokes to fresh concrete, by lifting and dropping the pipe every time. With this, the concrete is set well and creates corrugations on the surface, increasing the friction with soil. These piles are generally 300 to 400mm diameter and carry 80 to 100-ton load each. Driven cast *in situ* piles are common for industrial foundations where there is no rock and only friction pile seems feasible.

These piles have following limitations and better than cast *in situ* bored piles

- Can be driven up to a limited depth
- Not possible to drive in rock
- Cannot be used in marine construction

In these piles as pile is driven soil on sides get compressed thereby improving soil bearing capacity while in bored piles this advantage is not available and there are always dangers of side collapse.

8.10.5 Cast in situ Bored Piles

Cast *in situ* bored piles are alternative to driven cast *in situ* piles and are used when depth of pile is more or pile needing rock socketing.

Cast *in situ* bored piles are very common for cohesive soils with or without rocks at a reasonable depth. In case the rock is available, the rock is chiseled out up to a depth of three times dia of pile. This becomes an integral part of a pile termed as "socketing" in the rock. These piles are load bearing, in addition to frictional resistance from the rock and the soil around. In case the rock is not available, the pile has to depend only on the frictional resistance from the soil around. A hole of the size of pile is bored in the ground up to the designed depth by a pile boring machine, which could be a tripod or a massive piling rig. First, an auger is lowered in the hole with hydraulic pressure. This cuts the soil at the top of the hole. A bailer is lowered in the hole to take out all the cuttings of soil. There is another tool, called a chisel, to break rock if necessary.

This process is continued till the desired depth is achieved. Generally, soil around the hole is stable and does not collapse, if it keeps collapsing, it would increase the hole size or collapse could close the hole for pile, if small collapses, it may fall on top of the bailer or auger and they could get stuck in the hole. Or collapse while concreting makes pile useless. In such circumstances a solution of Bentonite powder mixed with water in right proportion is used. Density of this solution is kept more than soil in the pile and hence it keeps pushing the soil before it collapses in the hole. Bentonite is a natural product of soil, when mixed with water it becomes liquid thick solution having density more than compacted soil. All the operations of pile-making would continue as usual through the Bentonite solution. After boring to required depth is completed, hole is again cleaned with bailer. Reinforcement cage is lowered in the pile. Concreting is performed by Tremie process. while concreting this Bentonite will be replaced by concrete and come out of hole and preserved for next boring if possible. Cast *in situ* bored piles are extensively used in port jetties, approach bridges. Industrial foundations and high rise buildings

8.10.6 Continuous Flight Auger Piles (CFA Piles)

CFA piles are another means to make cast *in situ* bored piles. In this system the ground should be soft enough for an auger to drill full length of pile without difficulty. An auger with continuous spiral blades in full length is used to bore the pile. This auger is as per the size of the pile and is fitted on the piling rig, in a single length, longer than the depth of the pile. The auger has a continuous hole in the centre of about 100mm dia to feed the cement sand slurry for pile construction. Once the piling rig is placed in position and is ready to start, the auger is fitted to the piling rig. Fresh cement sand water mix is brought from the batching plant and the steel cage for the pile is ready at site. The cement sand slurry is fed to the concrete pump and the discharge pipe of the pump is connected to the auger on top. The pump and auger are flushed together with concrete at low pressure and slurry comes out from the bottom of the auger after flushing the system is completed and clear of any air pockets.

To prevent the slurry from falling off, a small plug is fixed on this bottom portion of the auger, the rig starts drilling by rotating the auger under the pressure from the rig. Within a few minutes, the hole is drilled, with Earth filled in all the gaps between the spiral blades and some cuttings of Earth coming out of hole. Once drilling up to required depth is completed, slurry pumping is started with pressure from the concrete pump through the hole in centre of auger. This slurry would push open the plug at bottom and start filling the drill hole, pushing the auger upwards, however little support in pulling auger upwards is also provided by drill rig. It is mandatory to make available full quantity of mortar as required at site before start drilling. Within a few minutes the whole hole is filled with cement sand mortar. Swiftly move the auger out and drop the pile cage in wet mortar. Cage will sink with its own weight in the fresh mortar and if necessary put a small pressure by an excavator bucket to push it down to near about bottom of pile.

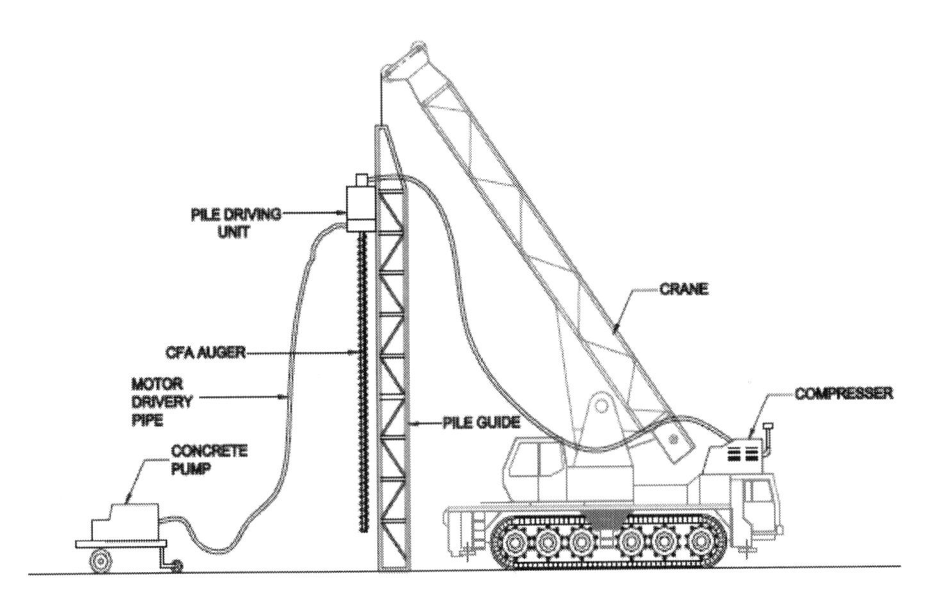

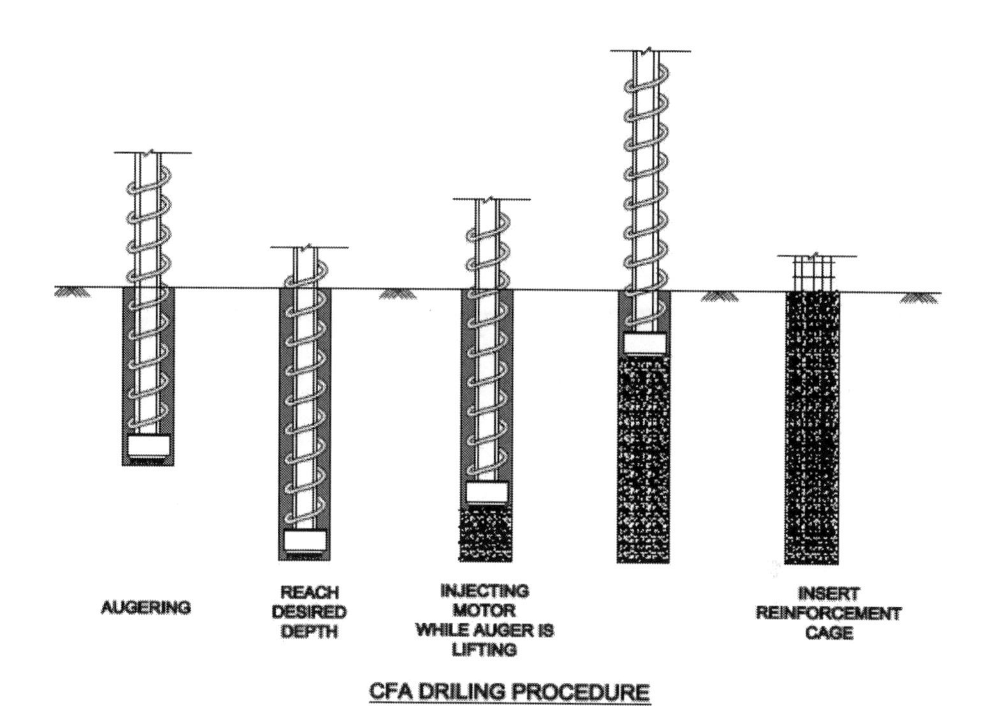

CFA DRILING PROCEDURE

This system has some advantages. There is hardly any sound compared to other piling operations so it is environment friendly. It is very fast, and since the auger is full of Earth around to size of hole, Bentonite is not necessary.

On the other hand, there are some disadvantages, too; It does not work in hard strata, where auger can't penetrate. Cement content is very high as, there is very high water content and no course aggregates. Further, the quality of concrete is often a challenge; it should not shrink or segregate. Due to high water content, avoiding segregation of sand and the consistent quality of concrete is difficult sometimes. To be safe, grout should be non-shrink quality by adding suitable additives. It is difficult to synchronize the pouring of concrete and lifting of auger. If a gap is created in between, Earth may collapse from sides and pile would not be effective below the level of Earth layer.

Necking of the concrete is more easily possible in CFA piles due to its speed. Care has to be taken in sequencing the drilling pattern. There should be at least two hours between two adjacent piles after initial set, which could be checked by tampering top of new pile.

CFA piles are preferred for smaller dia piles only Necking is reduction of diameter of pile in small distance due to a jerk or differential Earth pressure before concrete is become solid.

LEVEL 9

Projects Execution

We will now go through a few aspects of projects execution in major sectors of civil engineering.

9.1 Roads

After shelter, roads are one of the basic necessities for any civilization. It starts with footpaths and to date, we continue building super-express highways. The world has very efficient and effective road facilities, but in certain parts still, children have to walk two hours on hilly tracks through forests to go to school. On Sundays, people walk to church in the mornings, stay overnight and return home next day on foot. With the rise in population, there will be a rising need for extra roads all over the world in many categories. Some of these include well-dressed single lane road with Water Bond Macadam, up-gradation to two lanes, repairs to WBM roads and then sealing them with bitumen and stone chips to get better driving comforts, fewer pot holes and protections from rains. Finally, there are also four to eight lane roads with flexible pavements of asphalt concrete or cement concrete rigid pavements.

The first three categories are prevalent only in undeveloped countries mainly due to the lack of knowledge and resources. If the engineer tries to explain, the client agrees for the improvement of specifications of a project. For a new road, it is better and easier to take enough rights of way to avoid problems later while up-gradation. For any up-gradation of the road, an engineer should suggest the best alternative in quality and durability in the interest of the client in the given circumstances, including gradients, bridges and bends for the new road.

A Track to Graded Road and Beyond

A track for movement of trucks, carts and other heavy duty vehicles is created by the village good for fair weather season. With small rains this track becomes muddy and vehicle ae unable to move on such tracks till it is again dried up. Such roads are built with least public resistance and while taking care of sentiments of villagers regarding their houses, agriculture fields, wells, other water bodies, place of worship and other conveniences. This track is four to five meters wide.

Up-gradation of this track to road is a major activity for a good change for the local society. For the engineer it is important to understand that he is going to lay foundation of **the** road for this society and has to use his best innovative skills in engineering and persuasion to the society to provide them the best possible engineered road for ages to come. Further upgradations would be widening and improvement of quality of road structure with least change in levels and alignment. Resistance by the society at this stage would be least and its intensity would increase and may even multiply with age. Further development like building of houses market place, hospitals etc. would be along this road. World over any improvement done for the living of grand children is always welcome and supported hence development work has to be long sighted.

First land survey is carried out including ground levels and approximate road alignment of 12 meters' width is marked on the ground with minimum possible sharp bends, even at the cost of a few houses and dividing the agricultural property of individuals in two parts. Making new houses to replace those to be demolished should be included in budget of the road. Depending upon the cooperation by the society a final layout is agreed and again demarcated on ground with stones or wooden posts painted in a clearly visible color. There may be some deep ditches to be partially filled these ditches could be filled with support of good sloping bank or strong stone masonry on sides of road.

Cut and dispose of trees and bushes from the area demarcated for road including their stems and roots. Remove all organic soil from central seven meters of the road. Balance two and half meters on both sides is for drainage and a small footpath. Drains for rain water are built on both sides of the road at least 30 centimeters deep and sloping towards low area and ultimately discharge rain water in such low areas. These drains top should be at least two meters away from edge of road to save road collapsing in the drain or drain water seep through in the road foundation, hence minimum twelve meters' wide land is taken for the road including drain. It would be better if a few meters more width is available for the road. Some small culverts are also built for water to pass from one side to other and small bridges could be built even by timber as explained in this book.

Now cut and fill in thirty centimeter layers the central seven meters' patch with maximum possible compaction in decided gradient and bends to prepare a good sub-grade for the road. difficult to have a laboratory but still try to compact in a way that roller starts walking on fill material without leaving any marks on Earth surface. If sub-grade level is below the level of ground around, sides of the high area up to outer edge of drain should be cut in slope, say one horizontal to three vertical and some protection against collapse of excavated sides on road should be planned.

On this sub-grade first provide edge protection on both sides of road thirty to forty centimeters high with concrete curbstones or bricks or stone masonry, to protect the road edges against damages by traffic. A thirty to fifty centimeters thick layer of Water Bond Macadam is laid over the sub-grade with crushed aggregates, gravel, brick bats, slag or similar hard material locally available and compacted well. This material should be well graded to provide bonding between bigger stone pieces for a durable road. If felt necessary, a thin layer of soil is spread on top, watered to spread in small voids and compacted to create a good bonding in stone pieces. Centre of road should be 50 mm higher from sides for effective drainage.

Now, the area between the road and drain called shoulders should be dressed and compacted, by cutting and filling up to level of edge of road and connect to drain with a gentle slope towards the drain.

After a few years one more layer of 30 centimeters of stone aggregates free of silt and clay are spread and top, compacted to its maximum density. Now the road surface is cleaned with strong air jet or hard wire brush to clean the surface free of fine material. and then grouted with hot bitumen spread by a sprinkler in a layer of about three mm in dry condition of road, more the heat is better, it would be soaked and filled in the voids in upper layer of the road. After at least six hours spread a thin layer of aggregates 3 mm to 5mm size, and rolled with a road roller. This is called sealing of the road. bitumen will give binding to stone pieces in upper portion of the road.

Quantity of bitumen is to be adjusted at site and hence, contractor is paid for consumption of bitumen on weight of bitumen actually used on site. If bitumen is less, pieces of aggregates on top would slip out with friction provided by moving traffic. If too much bitumen is used it will spread on road surface in thick layer and make the road slippery for traffic.

Maintenance of Road

If properly done this road after bitumen sealing could last more than fifteen years, but drainage is to be maintained and small pot holes to be repaired if at all they develop due to mishandling the road by digging holes for some social functions etc. or damage while repairing a heavy vehicle, or loose pockets left at the time of construction. In case of pot holes due to loose Earth pockets under the road, all the loose Earth should be removed and then backfilled with sand or gravel, compacted and then sealed

While repairing heavy vehicles like cranes, dozers etc. care should be taken to avoid oil dropping on road and put a few sleepers under heavy jacks of the equipment. Jacks may sink and oil may mix with bitumen and dilute it. Normal trucks can be repaired by providing jacks. Only oil spillage is to be taken care.

Upgrades

It is desired but hardly happen, unless specifically mentioned in contract to improve sharp bends, super elevation and gradient of the road wile further upgradations since these works cost extra money and hassles related with matching design alterations and land acquisition. A very strong engineer and administrator are required to work together to implement such good changes according to requirements of engineering norms.

9.1.1 New Roads

New roads are built in townships and industrial establishments by first preparation of sub-grades and then sub-base, base course and asphalt or cement concrete as per specifications provided.

Further, with increasing urbanization in all parts of world, roads confirming to standards of express highways and super-express highways are being built at about a few kilometers away from the existing town road network as a ring road and connected by feeder roads to the town roads. To avoid traffic congestion, one could exit to the ring road at the nearest point, move on the ring road and reenter at the convenient junction to go the place of choice. These junctions are well planned, minimum in number but are connected to all the main roads to the town. Vehicles can enter or exit without disturbing traffic by using the extra lanes provided on the road at these junctions.

Super-express highways are built to connect two important towns without disturbing existing urbanization for smooth and undisturbed by new incoming and outgoing traffic. These roads are on high banks to avoid cattle from nearby villages come on road and this also takes care of un even ground levels by keeping the road in desired levels and grades. Other roads crossings are done either by bridges or under passes.

On these roads, minimum speed is specified and all the vehicles have to be driven within the specified upper and lower speed limits. Slow moving traffic is discouraged. These roads have a few bends with proper super elevation and good visibility specified on bends with large breaking distances.

Crash barriers, steel channels fixed on both sides of road as required to stop vehicles going off the road and centre verge, the concrete divider between drive ways in two opposite directions of adequate strength are provided. Centre verge is provided with green plants to break light glare to drivers in night from headlights of vehicles coming from opposite side. Adequate parking bays, away from main road, fuel stations, first aid posts, restaurants and public utilities are provided after every half hour drive at least 200 meters away from road with proper entry and exit lanes.

The biggest challenge for these super-express highways is entry and exit to and from major habitations by upgrading the existing road system to provide matching services without traffic jams. As regards construction, there is nothing specific or new except the quality of construction should be good with stringent quality control. Pavers for the top surface course laying should have good accuracy to provide safety and driving comforts at the speed the road is designed for.

Hill Roads

For hill roads, gradients should be a vertical to fifteen horizontals, and in exceptional cases, one could go to one vertical to twelve horizontals, in a short distance of 100 meters. For highways and on hills, the maximum gradient could go up to one vertical to twenty to twenty-five horizontals. These guidelines and specifications can change from place to place.

First, the ideal route is marked on the drawing, with existing levels and levels desired after construction of the road. Detailed survey pillars are built on ground. A set of equipment as per following table in capacities and sizes as per availability and size of the job are sent to the highest points of the road to connect the lowest designed location, generally as per drawing.

- Bull dozer
- Excavator with chisel
- Excavator with heavy duty bucket
- A few dumper trucks
- A vibro roller
- And a motor grader if available

This exercise is repeated for all the hills or carried out simultaneously for a few hills, with similar sets of equipment for each hill to be taken up simultaneously.

The bull dozer first makes the way as close to the road alignment as possible coming down from higher elevation. The efficiency of the dozer is almost more than double if it pushes stones downwards. The excavator with a chisel would break hard rock pieces embedded in ground to make it level. The excavator with a heavy duty bucket and trucks excavates loose material, including broken rock pieces and fills in low areas. If possible grader helps in making uniform level of road. This is the rough alignment practically good for four-wheel drive vehicles. Physically, the entire area is approachable, however, maybe with a lot of constraints and safety issues. This is the time to think, how best and fast one could complete the road, including some modifications to the alignment mainly grades and bends as close to specifications as possible. Once the final decision is taken on the alignment, the road up to sub-grade level is prepared by excavations, filling and widening of road including stabilization of side slopes of deep excavations and fillings. Often, a retaining wall of concrete or a stone masonry is a good solution for widening the road on valley side. Simultaneously, do not forget the installation of effective and good drainage. Carry out further work as per specifications and directions. We are not deliberating on work above sub-grade, since it is vast and if done as per specifications, does not have serious challenges.

9.2 Bridges

Bridges are built in RCC, pre-stressed cement concrete and structural steel. These bridges are designed and built for different utilities such as roads and railway crossings over waterways of all kinds, bridges built for crossing railways over roads and vice versa, flyovers in congested areas to ease the movement of traffic, and grade separators built at important and congested roadway junctions. Bridges on **land** are very easy to design and build since the construction of foundations, the pillars and bridge girders, known as superstructures, does not have major challenges since every activity is done on ground. While doing so, it is important to address logistics and safety, construction of foundations, sub-structure and piers construction and superstructure bridge girders construction.

9.2.1 Logistics and Safety

Mostly, road bridges on land are built to provide better traffic comfort on the existing road. Traffic management, while an effort to get good progress of work, is a major constraint. The following things are necessary. It is important to decide the minimum and safe working area required for work on the existing road. This area should be barricaded after seeing how much maximum area is available for the flow of traffic. Additional diversion roads may be made if needed. Enough traffic signals are provided to guide the traffic and drivers should be told well in advance that there is a diversion ahead. Traffic signals should be illuminated at night. It is important to carry out the maximum possible work at night and on holidays. It is also important to have a separate yard for storage of materials and preparatory works including casting of bridge precast girders. Enough safety awareness should be displayed at site. Stop the traffic while handling heavy objects like bridge girders with cranes on the site. Provide uniforms for all workers for easy identification and make it a point to keep the site clean, specifically roadways around the site.

9.2.2 Foundations

Foundations of the bridge are most expensive and load bearing structure. The designer first designs the beams in the superstructure depending upon availability of handling gear, logistics and then foundations. These foundations are generally pile foundations. Driven piles are avoided in towns due to high noise pollution. Bored cast *in situ* piles are preferred and driven for bridge foundations as per layout and sizes provided in drawings. Once piles are completed for a foundation, top one meter or more, being bad concrete mixed with soil is discarded and broken, called pile chipping and then foundation in RCC is casted first and then pier.

9.2.3 Sub-structure: Piers, Abutments and Approaches

Bridge piers are the part of structure seen by public on daily basis, besides being an important part of bridge. If the number of piers are more than six, steel forms properly designed are provided, fabricated by skilled fabricators. The entire pier is cast in a single concreting operation if possible. Form work and its supporting system has to be fixed properly as per design and drawings. Reinforcement bars are placed as per the drawing with proper spacing, laps and alignments. Special care to cover blocks, to be of good quality and placed properly. Even binding wires ends are bent inside keeping safe cover to concrete. Foundations are buried in the ground and only piers and superstructure are seen by public to comment on workman ship of the contractor in addition these are exposed to wear and tear by usage and atmospheric reactions, hence quality should be of prime importance. Further, these structures are designed considering good workman ship and very less room for mistakes is provided. If one drives along a bridge, he may see brown water marks on the structure. These are the locations where water would have entered concrete and steel bars started rusting. Brown color is rust and reduces life of the bridge.

Example

Thane Creek Bridge connecting Mumbai and Vashi, a bridge 1836 meters long with 39 spans was opened for traffic in 1973 but two years later discovered that in bottom of pre-stressed beams rusting has started endangering safety of bridge. Very soon temporary external support measures by tying with heavy steel cables were done for the entire bridge and another bridge was built hurriedly, before this bridge could be discarded. This was all may be due to either too much congested pre-stressing cables or careless concreting of the beams or inadequate cover to cables. Just for a mistake of a junior engineer supervising beams construction, contractor and Project Manager got a bad name and this monument is still standing without any use as a national waste.

Therefore, a lot of care has to be taken while working on these delicate structures in placing formwork, re-bars, pre-stressing cables and concreting. If there is a mistake, found even one hour after concreting, the shutters could be opened and concrete could be washed away or max the particular beam is rejected. The Project Manager should also visit site and inspect once a while, quality of work. It is preferred that after concreting is completed; an engineer goes to the structure and taps with a small hammer at many places thoroughly. If honeycomb is there it will give a louder metallic sound. Immediately open forms in the affected area and repair before concrete is set. As a rule, two engineers should check and sign the concrete pour card. On top of piers neoprene bearing pads are fixed at proper level and locations. Bridge girders sit on these pads for free lateral movement due to thermal expansions and contraction.

9.3 Construction of Bridges for Roads and Railways on Big Rivers

Nature does not tolerate any interference in its shape, size or process and very strongly reacts to such interferences. One should not play with nature. One cannot go against nature, it will destroy but if one cooperates, nature will help. This philosophy would be experienced by an engineer while working for mega projects and will have to always work keeping nature in mind. The author was taught this on his first job which was one of the biggest bridges of that time in India. He has always been successful due to his nature friendly planning and behavior on all of his projects.

9.3.1 Model Studies

Model studies are conducted for design of big bridges, dams, sea ports, airports and all other projects involving natural water bodies, air and atmosphere. Model studies ascertain how best the project structures could be nature friendly involving interference with natural sources like water and wind which are powerful to resist any interference in their movements. There should be least resistance to structure from nature like wind, water and soil. How the project or an object in the project should behave to take maximum support and least resistance from nature and ultimately what are the unavoidable resistances and natural supports. Their valuation is carried out for taking care in design and execution of the project.

In a sail boat it is the skill of the sailor how fast with least jerks and bumps he could travel a long distance without any engine or by his own physical effort in the boat. He takes full support of waves, water current and wind. Even shapes of boats, cars and aircraft are designed nature friendly by researches in aerodynamics.

Hence model studies are conducted for each project involving interface with nature.

There are a few laboratories around the world, having models of topography of land mass around for major rivers and sea coastal areas and bed profile plotted and built to scale on ground. They also maintain data of air and water current and their direction for each month of the year and keep on updating such information. They have equipment and resources to produce water flow with currents, air movement etc. according to the scale of model to induce the same environment as on proposed site of the structure to be built. This data is upgraded regularly.

Feasibility Study Process

To conduct feasibility study of a project, there are a few laboratories in the world, one in Pune India is **Central water and power research station** covering India and a few more countries. Models of marine structures to the same scales as of ground profile in laboratory are fabricated in different alternatives of shapes and sizes. These models of the structures to be built are placed in the desired location on the model of the laboratory in different angles and directions one by one. Stipulated air and water currents are created around the model as existing in the location of the project. Different models are tried with different angles and adjustments of locations to find out which is most favored by nature, means which one would have least resistance and maximum support from nature and how to plan its orientation. Northings and Easting's at critical points are recorded with details of air and water turbulences and other important data sent to design engineer for further work.

Now the same exercise is done by computers but still results are rechecked on physical model if possible. A sea port was being built on banks of a river and as a common sense was felt good to make the port parallel to shore line. But the model studies recommended the port to be at three degrees' angle to shoreline. And it worked very well. Bangla Desha country full of big rivers has a dedicated computerized laboratory to monitor behavior of rivers and suggest safety measures against erosion of shore lines and provide safety signals to the government on regular basis. In Bangla Desh, as a rule no new important structure is built even on land near to river banks without clearance of this department. Once a thermal power station was planned quite away from banks of Brahmaputra River, by the side of a bridge but had to be shifted perforce on recommendations of this department.

For airports construction, orientation of runway should be necessarily decided by model studies, otherwise slight push by heavy winds, aircraft may lose the track while landing or speeding on runway for takeoff.

9.3.2 Guide Bunds

Guide bunds are two large bunds built on either side, parallel to the flow of river with rock fill as per profile and layouts proposed by model studies. The bridge is built within these bunds and river should flows between these bunds at that location. further there are big masses of land, behind these bunds and get flooded with river back waters during floods. Separate embankments with high banks are built on both sides of the bridge to connect to the road with these guide bunds. A few drainage culverts are provided in these bunds for water to travel one side to other. Such arrangements are witnessed on all the major river bridges around the world.

9.3.3 Pier Shapes and Scour Around

Pier shapes should be as comfortable as possible for the river water to pass around. Piers are made in a length parallel to the current, with rounded or parabolic ends. Due to their obstruction, water gets agitated and the soil on the river bed erodes up to a substantial depth around piers and well or piles foundations in high currents and are filled back when water current is reduced. This is called scouring around the piers. It is very important information for the designer calculated on the basis of model studies. Piers and foundations are designed for maximum possible scour depth.

Take a glass of water and put on the stable table. Put a rod in the water vertically in centre of the glass. Rotate the rod fast, vertically creating agitation in water. Water level around the rod will go down while agitating and back to normal on stopping agitation.

9.4 Foundations for Marine Structures

For bridges and other marine structures like sea ports, foundations are difficult and time consuming to build – but they are important parts of the structure. Once the foundation work is completed, the remaining work though complicated, it offers relief that we can finish the bridge quickly since the dependence on nature is reduced and is only a matter of putting resources together. There is no shortcuts or compromises available to the engineer to complete the foundation early or to terminate it at shallow levels. Foundations have to be built better than specified, not just as specified. If a pile is being driven, after the task is completed as per specifications, a few extra blows of the hammer are delivered on pile, just to be safer. Bridges are built with a minimum number of foundations to look good and to be economical and faster. Bridge foundations are in the types of well foundations, bored cast *in situ* piles foundations, driven steel pipe foundations, driven sheet piles foundations and precast caisson foundations

9.4.1 Well Foundations

Well foundation is a big cylindrical mass of concrete sunk in the river bed even to the extent of more than one hundred meters to support the bridge, this mass is kept hollow for construction purposes and when construction process of building up and sinking in ground is completed, its open bottom is closed by a heavy concrete pug and rest of hollow portion is filled with sand or water as per design engineer's instruction. This well foundation by all means is good and safe foundation of heavy bridges. Top of the foundation is closed by a thick concrete slab called well cap. On top of well cap, pier is built to support the bridge.

Well foundations are generally provided in strata with deep sand and clay layers and practically no rock up to founding levels. Wells are used in different shapes like **O, B, 8, square, octagonal** etc., vertical cylinders of RCC concrete open at bottom. Further bottom of the well tapers to a conical shape as shown in the drawing with a steel shoe called cutting edge. During dry season some area of river bed becomes dry and this is best to start well foundations at all possible dry locations. Even little water, also could be made dry by filling with loose Earth just for purpose to place cutting edge and a few lifts of concrete.

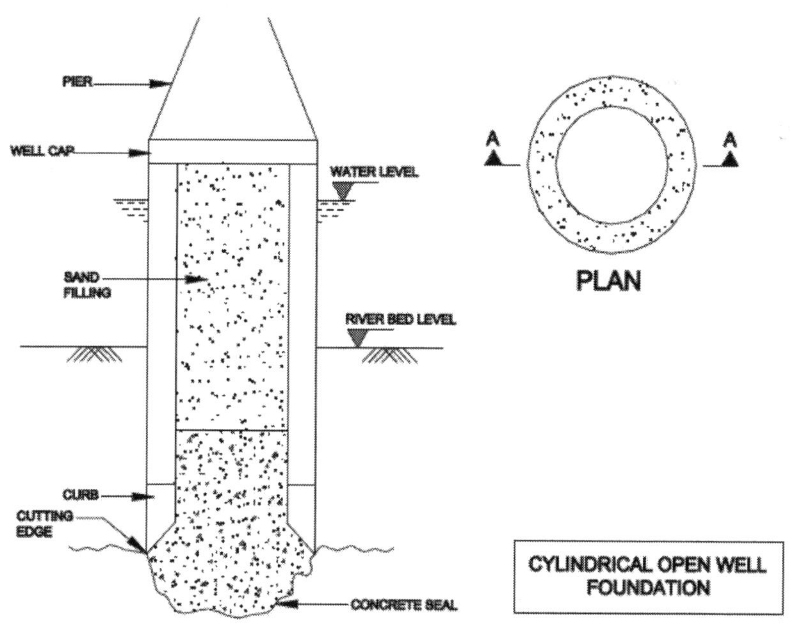

First, the cutting edge is placed correctly on river bed in the location of a foundation. Re-bars and formwork are fixed as per drawing and concreted up to curb height. The concrete of the wall is continued up to a reasonable height, say five meters. A grab is lowered in the well, the centre is attached to a crane and Earth is removed from the well. As the Earth is removed and a pit is created inside the well, well would start sinking in the ground due to its own weight. The speed of sinking is fast in start with but slowly reduced due to friction of ground on wall. Efforts are made to have all the wells, at dry locations sunk in ground at least equal to dia of well and walls are concreted up to well above the high water level before floods. Further concreting is done by concrete mixer and related infrastructure mounted on a floating barge, and well sinking by a barge mounted crane. Sinking of the well and concreting of walls continue simultaneously till level of cutting edge is reached desired founding level/final designed level. However, if not completed, sinking work is expedited in next dry season to again carry out maximum work before floods.

Well Gets Stuck at any depth

First check if enough of a pit is created inside the well, but never deeper than the free height of the well above the ground. Move all the equipment away, say a minimum of twenty meters and start dewatering the well with a power full dewatering pump, as the water goes down, external pressure on Earth inside of well will be far less than outside load since burden of water column is reduced inside the well. Ones it comes to breaking point outside Earth will blow inside and well will have practically no friction for a moment and will go down with a jerk and Earth on sides will fill up in the well, and removed after it is stabilized. Sometimes surface charge of a small explosive is provided in the well to give a shake and this also gives some good results, just one or two sticks of gelatin with a detonator is enough but too much explosives may crack the well, which is an expensive damage and client may ask to fill the whole well with good concrete up to the top level of crack. Mostly all wells need extra weight say concrete precast blocks, called cantiledge at time nearer to final sinking. Always, skin friction provided by Earth to external walls of the well has to be far less than total weight of well including cantiledge load for sinking of well.

Often in the above process of forced sinking and even otherwise, the well tilts on one side. To repair this tilt, grabbing should be done from only higher side and loading of extra cantiledge are put on top of higher side of well. Before starting grabbing inside one should check if grab bucket would really reach the point you need and answer is often no. in such cases, some grabbing should be done from outside the well in higher portion. After well sinking is completed up to required level, well walls called staining are raised to level as shown on drawings by concreting and loose Earth inside the well is removed and cleaned. At bottom of well a concrete plug is installed up to a level, little above curb or as shown in drawing. This concrete is poured in the well either by Colcrete or Tremie concrete as per convenience. The balance portion of the well is either left filled with water or sand filled in it up to bottom of well cap level as per drawing. On top of fill, well cap in RCC is installed with dowel bars for pier.

9.4.2 Bored Cast in situ Piles in Wet Conditions

Bored cast *in situ* piles are used for bridge foundations in a cluster of a few piles as shown in the drawing. A jack up rig is a barge with four spuds. Spuds are long fabricated structures fitted in strong casings on four corners of barge. these spuds move up and down in their casing with help of hydraulic winches. Barge is brought to the location, spuds are lowered and barge is lifted on these spuds by hydraulic winches. Rig becomes a stable work platform in deep waters at the location of foundation. Piling work is carried out by the equipment mounted on this platform. First, a steel casing pipe with inner diameter slightly more than the pile is lowered up to river bed and hammered to become stable and properly fixed in the bed This pipe works as a casing for the new pile. Cast *in situ* bored pile is built through this casing and is concreted up to pile cap level. Wait for an hour, cut the empty casing pipe above the pile cap level and move to the next location. For concrete and other materials supply, a separate ordinary barge parked near the jack up rig is used. Depending on the size of the rig and the foundation, the piling operation for one foundation can be done in one or multiple sittings of the jack up rig. A combined pile cap covering all the piles is constructed on top of these piles. Pier and other structures are built as per drawing above pile cap.

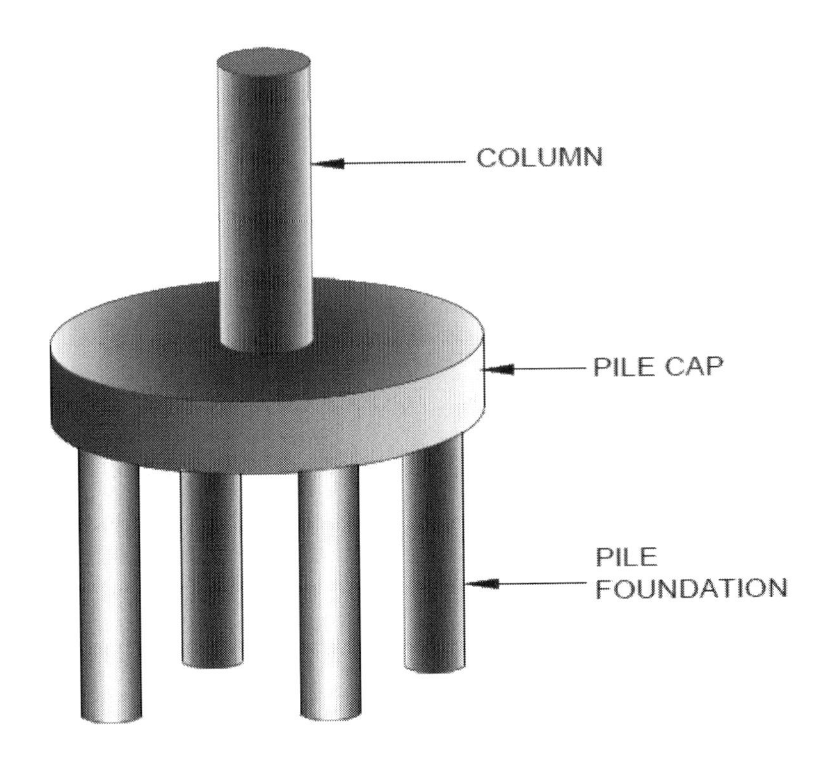

9.4.3 Driven Steel Pipe Foundations

Bored cast *in situ* piles have their limitations of depth and time for driving each pile. For deeper foundations thicker steel pipes are driven at location by a crane and powerful vibro hammer. After driving these piles, empty space within the pile is filled with concrete. Trimmed at the level of pile cap and pile cap is constructed. This operation is carried out by a crane mounted on a barge with strong mooring winches.

9.4.4 Sheet Pile Foundations

Sheet piles are available in many shapes and sizes. These piles have the provision of interlocking. After driving the first pile by a strong vibro hammer, the edge of the second pile slides in the groove provided in every pile, giving it a strong joint. This way, all piles in the foundation are driven by hammering, making a continuous wall. These piles could be driven in water by a crane on a normal barge and does not necessarily need a spud/

jack up barge. These piles are more useful on rocky strata. Drive all piles in the shape of a large foundation, O, B, or 8 etc. All piles are driven to refusal. The foundation is filled with water or sand up to bottom of pile cap as directed by engineer. On top of this sand layer, which would be the bottom of the pile cap, a big raft as per drawing embedding top of all the sheet piles is concreted.

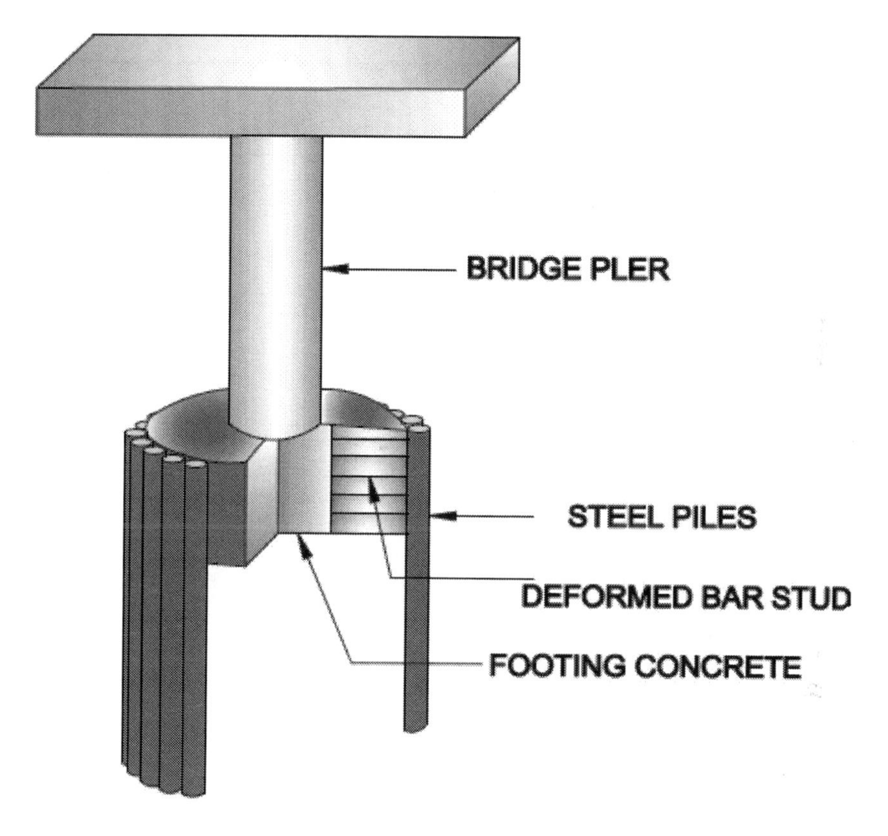

Sheet piles are also used as a temporary measure for building a well foundation. First piles are driven on the river bed in form of an enclosure bigger than size of the well. After completing the piling just enough to stand of their own on river bed as a box, fill inside the piles with stand and build well using the sand bed as a base. After the well goes down into the river bed substantially, these sheet piles are removed for further use. This temporary platform is called a cofferdam. Care has to be taken that these sheet piles close in the end of the circumference of foundation. The surveyor should be vigilant. Last few piles are assembled together and close the loop with interlock and then driven in stages one after another.

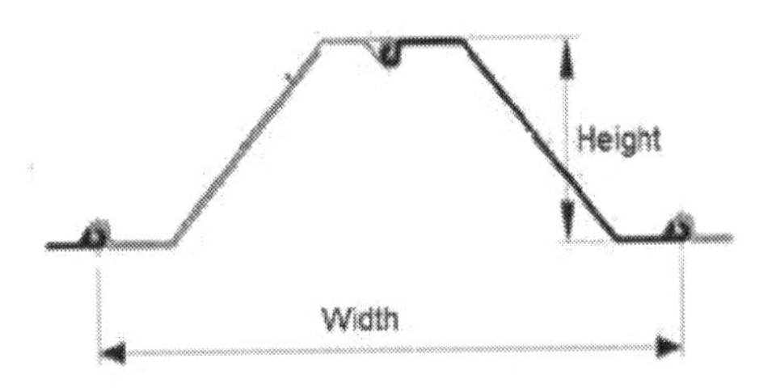

Z 'SHAPED SHEET PILES

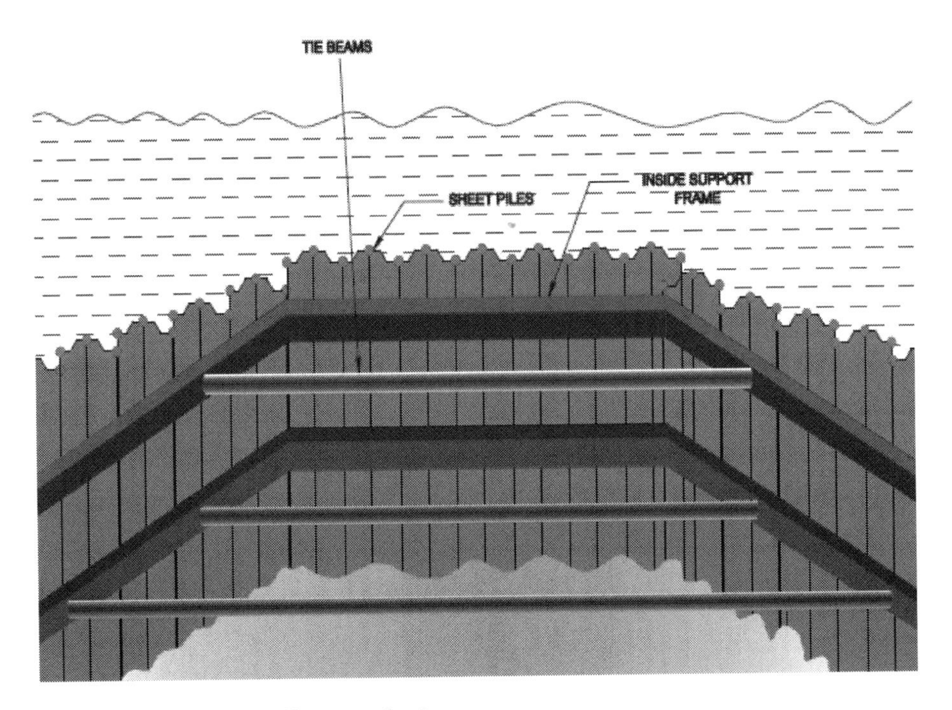

Cofferdam for Well Foundations

View of half of cofferdam or sheet pile foundation with driven sheet piles and support structure which is necessary until substantial filling inside the cofferdam is completed. Since all piles are inter connected, after filling they would not fall out in normal circumstances. Internal supports are removed as sand filling is progressed.

9.4.5 Precast Caisson Foundations

In cases where the river bed is strong enough to take load of the structure through foundation just sitting on river bed and such strata continues to substantial and safe depth, concrete foundations called precast caissons sitting on the river bed are planned. If the bed is sloping at the location of foundation, adequate measures are taken to avoid any movemement of caisson after final sitting on the bed.

The foundation area is cleaned by a crane and grab. Some differences like pits in beds are leveled with a rock mattress by dumping stones from the barge carefully. Sounding(depth of water at location is measured by droping a chain or tape upto rever bed level), are taken while dumping stones and a level base is created for a caisson to sit. The cassion in a floating condition is brought to the site and at least twice, it is made to sit on the bed's foundation. With this, the stones are compacted well. Repair the level of the foundation by putting some stones in the lower areas if necessary. Now, bring the caisson to the location and sink it properly in the designated location. After sinking is approved, the caisson is filled with sand and the cap is concreted to build a pier on top of this foundation. This cassion should not tilt or move when it is subject to loads from a bridge. For this,make the caisson big enough to take the loads due to its weight, or grout the stones in the mattress under the caisson, In worst cases rock anchors are installed as additional safety.

9.4.5.1 Building a Caisson

Precast concrete caissons are used in foundations for bridges, sea ports and other works like heavy power transmission lines,jack wells in rivers for drawing water for municipalities etc.. They are made as hollow concrete blocks which have closed bottoms with thick concrete slabs. All the outer vertical walls and some inner walls are built as required. The top remains open. Caissons are built either by conventional methods of RCC construction or by slipforming, depending on the magnitude of the job and availabelity of slipforming equipment. These are heavy precast structures and are thus partly prepared on launching platforms, and partly while floating in water. The height of the Caisson building on launching platform should be a little more than adequate for the safe launching of cassion. Two questions are asked by a layman

How does the concrete structure float in water? If the overall volume of a structure in cubic meters is more than weight of structure in tons, structure would float in the water. Water is having a density of one ton per cubic meter. In other words, if volume of the water in cubic meters or weight of water in tons, displaced by a structure is more than weight of structure, the structure would float. This universal law is applicable to ships made of steel, boats and all the floating structures. These structures are made hollow to increase the volume of water displaced. Structure floats with a portion of it in water, which represents weight of water displaced equal to weight of structure. Lighter the structure, less water is required to float. Floor area multiply with depth of structure in water while floating is the volume of water displaced. this volume multiplied with density of water equals total weight of structure.

Why are concrete caissons made in two parts? Caissons are first made on launching platform or in a dry dock in a minimum safe height/ weight for easy to handle and less water depth is required for launching and the balance is built up in floating condition in deep waters. Height of the caisson at the time of launching should not be so low that it may sink. Buoyancy created by the caisson has to also break bonding between bottom of caisson and dry dock floor. While launching on tilting platform, caisson first goes a bit deeper in water due to its speed in slope and then comes up. The buoyancy pushes up the front portion to make it level and float.

9.4.5.2 Flat Launching Platforms for Caissons

Flat Launching platforms are made on bank of river or sea shore with a slipway for launching in water at an angle about 30 degrees. Flat portion is bigger than base of caisson. Entire platform and slipway are covered by wooden floor. Floor is given a thickcover say 5 mm of yellow greese and then covered with a HDPE film. On this base slab of the caisson is concreted. Give about one foot push to the slab towards the slipway to break the bond if any betwwen slab and top of launching platform. Now concrete the caisson by slipform method of concreting to the safe minimum height. Around ten days later when concrete is strong enough, push the caisson by special jacks which could push such a mass in a single non-stop operation little more than half width of caisson. These jacks are patented and suplied by many vendors and without such jacks, this system should not be tried. These jacks are contineous climbing jacks climb on 75mm x75mm steel rods and many jacks work togheter.

9.4.5.3 Tilting Type Launching Platforms

This launching platform is made of steel, which is slightly bigger than the caission base. The platform is pivoted in the centre and is hinged on a concrete foundation as shown in the drawing. The caisson is built on the platform and when it is ready to be launched, the backside is lifted by a hydraulic jack as shown in the drawing. The caission slips into the water. Care must be taken to see that the centreline of the caisson is slightly away from the river side from centre line of the platform. This gap is retained in order to put more weight on the landside of the launching platform. Otherwise, there is a possibelity of the platform tilting on its own, and not only will the half-finished cassion would be launched, but serious injuries to workers on the platform. In the diagram, a concrete block is shown on the riverside, which should be strong enough to take the sudden impact load of the caisson and the platform. Generally, this block is built as a pile cap on the row of piles. This block ristricts the angle of the caission launch. Without this block, the angle of launching will increase and the caisson may topple. While launching the platform sits on this block and should not tilt more than twenty to thirty degrees.

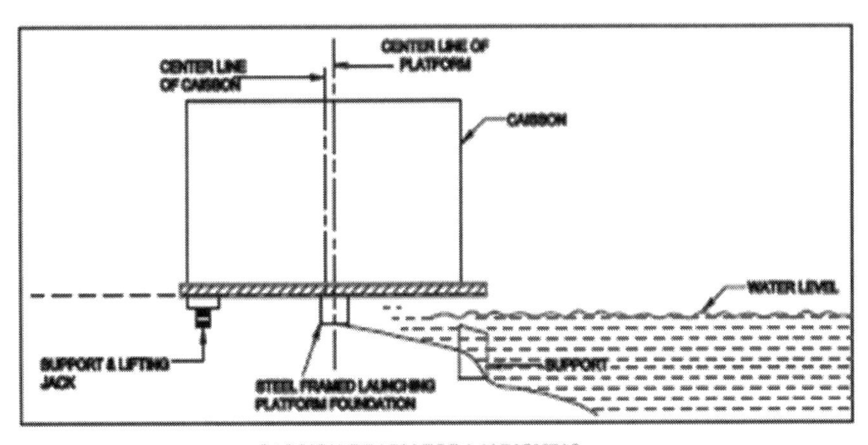

CASSION READY FOR LAUNCHING

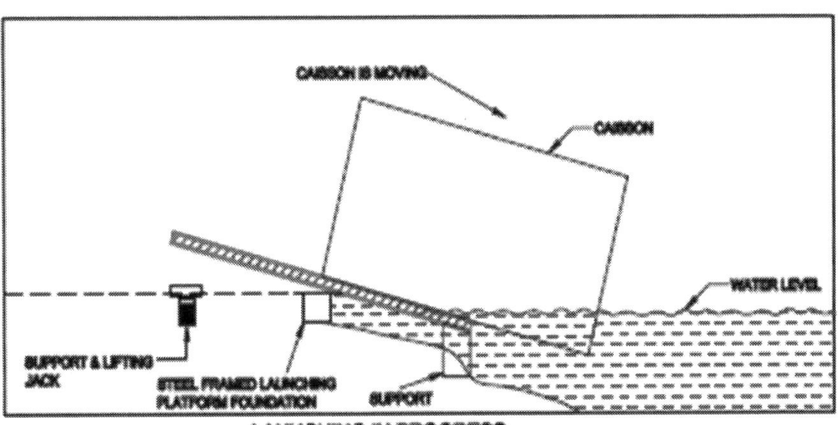

LAUNCHING IN PROGRESS

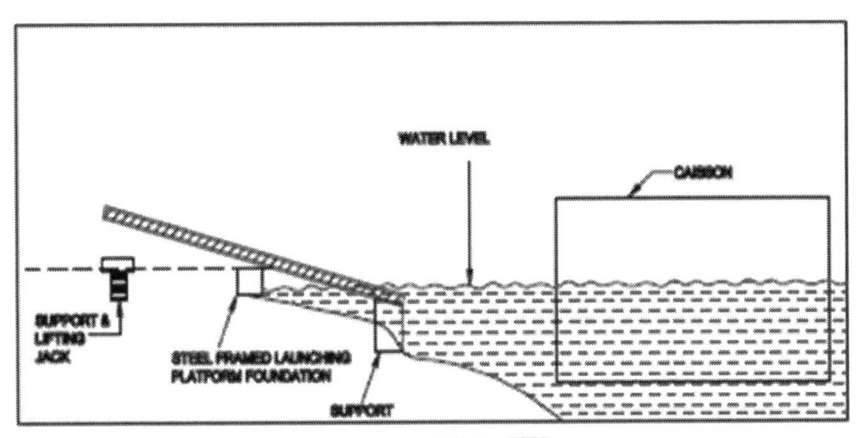

LAUNCHING IN COMPLETED

9.4.5.4 Dry Dock

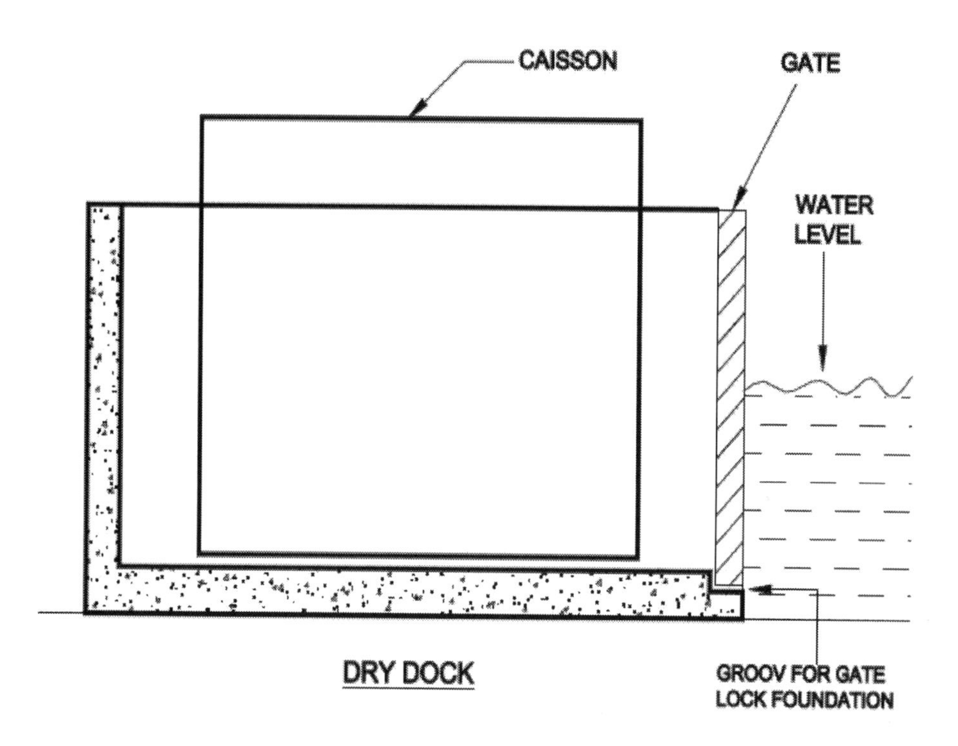

A temporary dry dock could be made at site for building caissons. A site is selected where the water in the river is deep enough to keep the caisson after first stage concreting floating comfortably. Depending on the nature of the soil, about ten meters or more from the edge of the high bank, dig a pit to a depth at which the caisson will flaot, with at least half-a-meter of clearence between casisson bottom and bed level of the finished dry dock. That is, while the caisson is afloat, there should be at least half-a-meter gap between the bottom of the caisson and the base of the dry dock. Keep at least two meters on all sides as a working clearance. Make a slab of approcimately 30 cm in thickness on well compacted and a level base. The slab should have nominal re-bars and provide dowels for wall concreting. They need not be vertical, but wall should not collapse when filled with water. On the riverside, prepre a u-shaped structure on the bottom slab to house the dry dock gate. Vertical offsets on both walls are provided in line with inside face of the concrete seal for gate in bottom slab. A gate is a floating steel stucture placed in front of a dry dock. Water is filled slowly in the gate and it would be guided to sit in the grove provide in base of dry dock. Both the top ends of the gate are pulled in by winches. Gate touches the vertical bar on both sides. Start dewatering the dry dock. Due to water pressure from out side, gate will firmly touch the concrete offset in walls. Better to provide a rubber seal on the areas of contact of the gate with dry dock. For opening the gate, dewater the gate, it will slowly come up and allow water to enter the dry dock. After the dry dock is constructed, the Earth bund in front should be removed. A water channel created for taking out the caisson from dry dock to deep waters in floating condition.

9.4.5.5 Caissons for Small Foundations

For construction of small structures in sea and deep rivers like jackwells for taking water from deep portion of river for drinking purpose, foundations for power transmission lines, intake structures for sea water needed for cooling of plant etc, caisson foundations are easiest and cheapest solutions. It is too expensive to make arrangements for building a few small caissons with eloborate arrangements deliberated earlier. One of the procidures for construction of these caissons could be as follows: Refer to drawing of the well foundation. Prepare a steel formwork for the well upto a few meters of curb of the well. Cutting edge of a small angle say 75mmx75mmx6mm is made in circular or rectangular shape as per drawing and placed on the banks of river. On this cutting edge build about two meters high shell of thin weldable steel plate say 3mm as minimum in the shape of curb. This structure is hollow shell with bottom open. Fix enough bracings to avoid bending while handeling.Drop in water with a crane, this structure will float. Concrete bottom of the structure in about 30 to 50 centimeter height provided the shell is still floating. This will provide good strength to the structure and stabelity while floating. Lift the steel form work further and keep on concreteing, with a care that the structure always remaied in floating condition to a comfotable level and deep enough water is availble all the way upto location of foundation. Now pull the floating structure to the place of final sitting and continue further concreting till it sits on the river bed. If desired you can dig inside the caission and sink it as per requirement. cassion is ready for further work as per choice and requirements. This system is also adopted for a few bridge well foundations in deep waters.

9.5 Piers

The piers of a bridge are simple structures. They are normal concrete colums in a particular shape that can be built with a normal process. Efforts should be made to complete the piers from one side in order to enable start of superstructure construction. With this progress of project gets improved. Care has to be taken for keeping top of piers at the same level of concrete. on these piers are are small padestals. Neoprine bearings are fixed on these padestals for movement of bridge girders due to heat expansion and contraction. Bridge girders are gently placed on these bearings. Earlier there were steel bearings, fixed and rolling. On one pier fixed bearing called rocker bearing was fixed and on other pier rollar bearing was used to be fixed. Neoprine bearings are steel plates molded in neoprine and all the bearing accomodate for movement of girders freely kept on them.

9.6 Superstructure

Superstructures of bridges are primarily beams between the piers and road surface above them are very delicate. Practically every bridge has some innovative design and construction technique to make the superstructure strong enough, sleak and good looking. Components of superstructure are to be handled carefully and with vigilance, as per the directions of the designers. Most new bridge superstuctures have a number of precast pre-stressed concrete elements that are erected by floating equipment or launching trusses walking on top of bridge or a combination of both. This equipment is made to order to the work specifications, or by modifying existing equipment if available. Most commonly, precast pre-stressed concrete beams that span between the piers are erected either by a launching truss, a floating crane or are brought to site on a barge in high tide and placed between the piers using tidal variations.

Launching Truss

Erection of a precast structure between two piers by a launching truss is shown in the following sketch. All precast beams are made in a yard on one end of the bridge, left side of first pier **A.**

Precast beams are brought from this side and keep on launching in their position. Construction of deck slab over the beams is taken concurrently and continued till end. In the drawing top slab is concreted between piers A to C and launching truss is moved to span C-D after deck slab concreting of spanB-C. Two front legs are anchored on pier **D**. Front legs of the truss have moved in air as a horizonaly hanging cantilever of truss suported on two set of legs on its back. But for taking load of beam its front legs have to firmly stand on next pier. Next set of beams are brought one by one to from back of the launching truss. The truss lifts the beam and hooks to two traulies. Traulies move on the truss forward over to the new location between the two piers and slowly and beam is lowered by hydraulic jacks in centre of bridge. Beam is pushed sideways to its permanent location, manually with help of jacks and then next beam is brought and placed till all the beams in this span are complited.deck slab is concreted and truss moves to next location.

Uptill about fifty years back snd jacke were used for lowering the beam slowly between piers. There used to be dry sand filled in jack. Open the hole at bottom of jack and start vibrator, slowly sand will drop off and beam is lowered.

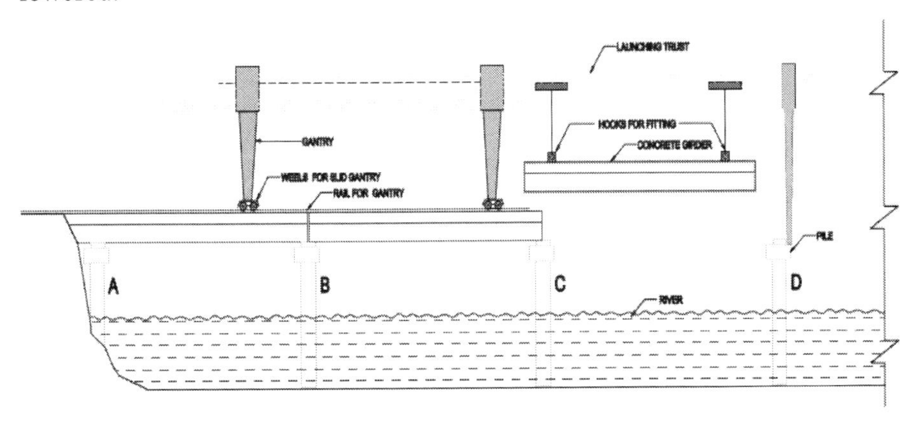

Floating Crane

Launching of a precat girder with help of a floating crane. Beam is brought on a floating crane barge to the site and slowly lowered in its position as shown in the sketch by the crane.One would notice in both the cases that selection of equipment for launching of beam is very much dependent on weight of the beams, the heavier the beam, heavier the equipment and the equipment also has some limitations, hence selection of length of beams and distance between the piers is selected considering the equipment and method of launching. This is possible only in pre-stressed concrete or steel structures due to their light weight.

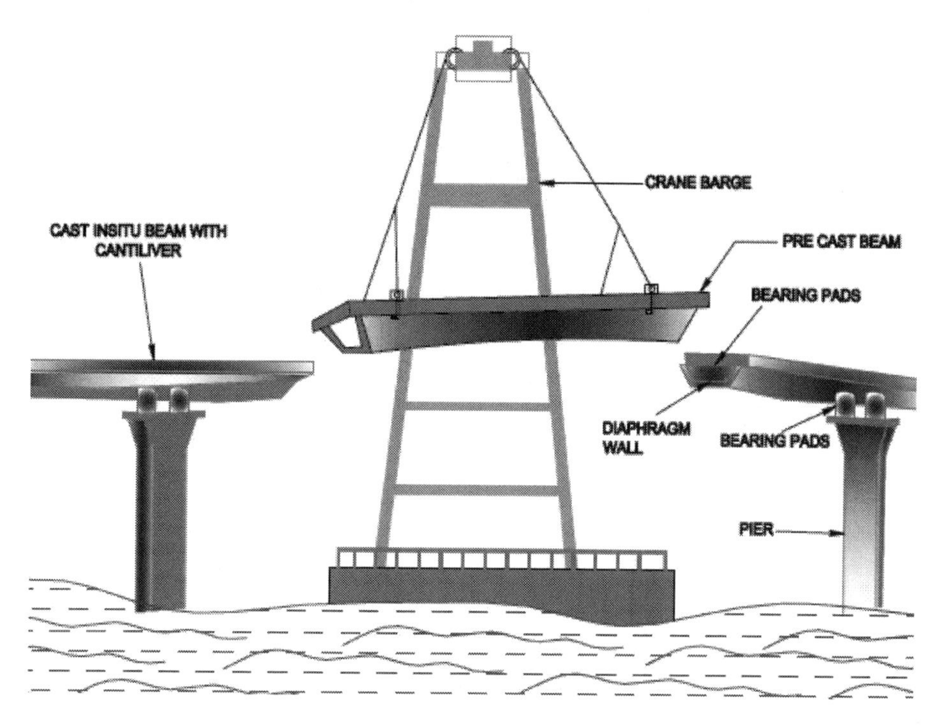

Insitu Concreting of Bridge Cantilevers

Cantilever portions of the bridge is concreted by slipform moving horizontally with combination of reinforced concrete (RCC) and pre-stressing arrangement. This is done by a special forwork fabricated to order as shown in the drawing, concreting is done steps by step in segments of about five meters. When concrete is strong enough to take its own weight, form work is moved to the next locationand fixed as hanging with ancor suports on its back. In the mean time previous concrete is strong enough with age to take load of the new segment. Concrete is designed to work as RCC concrete for its own weight and then pre-stressed to take aditional load of precast beam between the two cantilevers and traffic. Concreting cycle is roughly by, Dismental forms ten days after concreting. Next concrete twenty days after previous date of concreting. This way every twenty days move five meters.

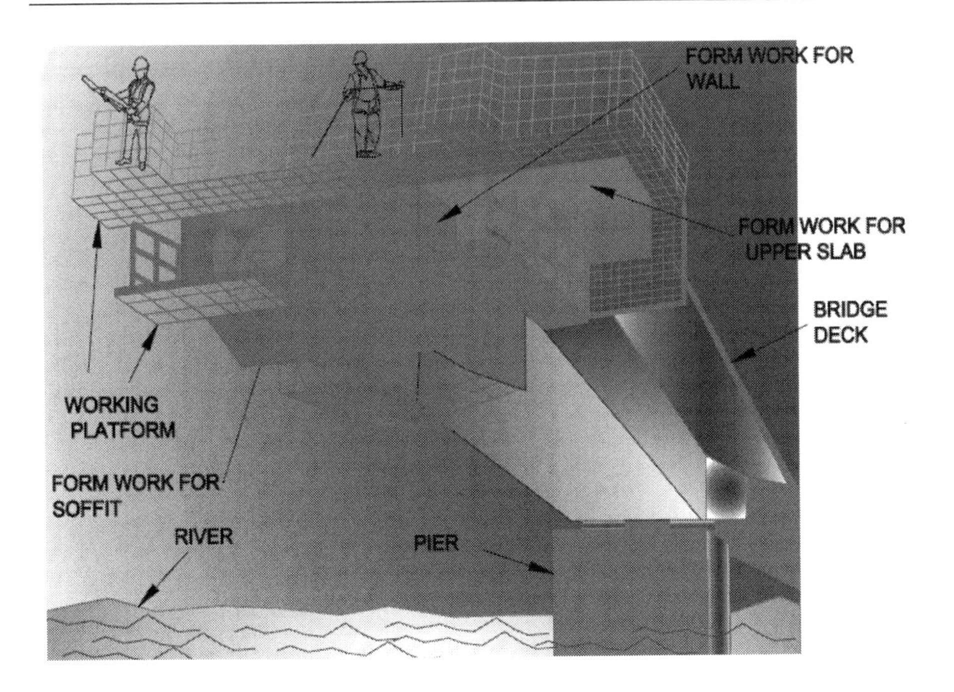

Cable Stayed Bridges

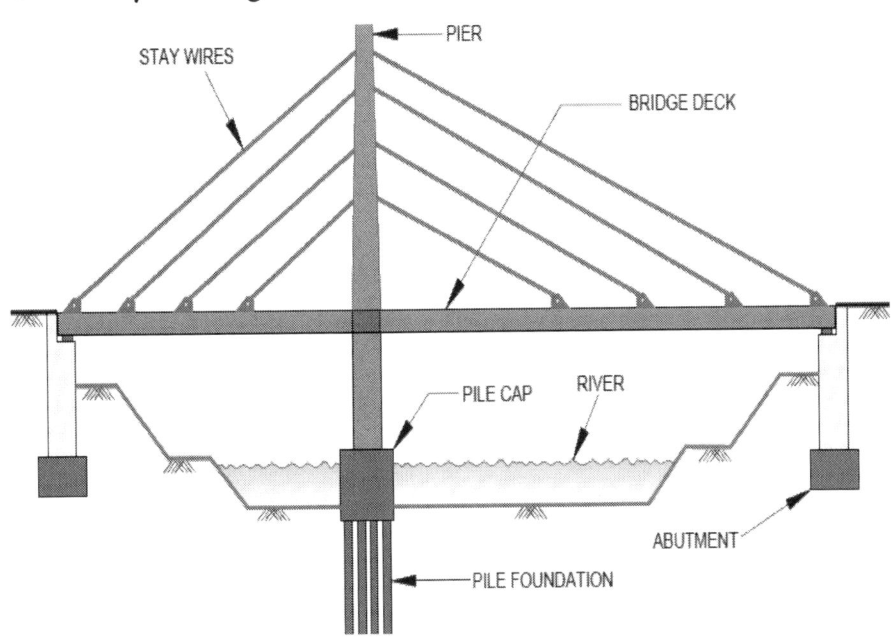

Precast elements are suported by cables that are connected to the pier. They interlock with each other and work as one unit after pre-stressing at different stages of construction. Care should be taken that the pier has equal load on both sides. In the above drawing, the pier is on one side of the river to allow big boats to move under the bridge. In such cases, the precast slabs on the shorter span should be heavier than the other side of the pier to balance the load on the pier. Else, the pier has to be designed to be strong enough to take a permanent bending the load on one side which is not desirable. In best situations to avoid unequal load on two sides of the pier and avoid good amount of bending stresses. In such ideal cases, pier is provided equidistant to both the abutments.

9.7 Dams

Dams are built across rivers to control floods and for hydro-power generation, irrigation, drinking and other water-related utilities. They control of water flow in areas in downstream of the dam and the supply water to cities and townships for different utilities by avoiding excessive discharge of good water in the sea.

A dam is built across a river at a suitable location which is primarily decided after an elaborate investigation is carried out. A dam is located between two hills on either side of a river. For example, if the dam height is twenty meters, if the river is blocked at the chosen location, these hills on two sides of river will hold the water up to that height, without water spilling over to the other side. Catchment areas should also hold the water up to height of the dam without any exception. Otherwise, the level to which the catchment area would safely hold water would be the level of the dam and the height is reduced as required to building the dam in that location. The total area of catchment also called submergence is calculated and plotted on a drawing with, locations and existing ground levels of houses, market, schools, roads and forest are marked to sum up the losses to the structures and utilities due to submergence in stages up to different levels. If submergence height is reduced, some structures etc., at higher levels would be saved. That would also reduce the height of dam. An elaborate study is carried out for the feasibility of building a dam at the proposed location. This is done with the given parameters, i.e. whether it is technically, socially and commercially viable or not. If it is found feasible and the height of the dam is decided, a detailed design process begins. Work proceeds as per the prevailing practices of approvals of cost, social clearances and analysis of costs against benefits and losses called **cost benefit analysis**. If substantial benefit is seen than only project is pursued further.

9.7.1 Construction of a Dam

Usually, in dam construction, huge quantities of different activities of work are executed. Construction methodology and planning is discussed in the earlier chapters of the book. Mobilization of resources including men, material and construction machinery are required in good numbers and their selection, timely mobilization, costs, finance, planning and logistics are very important. A fairly big organization is necessary for construction of medium to major size of dams.

Major challenges in construction are the building of foundations and river training. Usually, the length of the dam is a few hundred meters. Ground levels are the lowest in the river bed generally less than 50 meters wide in dry season and keep rising on both sides. A flat and slightly higher than the river bed portion of land is located in the section of the dam and such an area is selected for the river diversion channel. The river flow will be diverted through this area after some time, and will flow until the completion of the dam. The diversion channel should be of a minimum width for water to pass comfortably in about one-meter depth in the dry season. If possible, the bottom of the channel can be raised but may be kept lower to the top of the water in the river in the dry season. This area should be selected carefully in the concrete portion of the dam and advised to the design and planning engineers to plan their work accordingly.

Next, a few holes are drilled along the length of the dam to check the level of the rock and the ground water in these locations. First, build the foundations of the dam in a dry condition where the ground water level is lower than the foundation level and raise the dam by a few meters above ground level.

For the foundations where ground water is slightly above the founding level, depending on the ground strata and water level differences, install a few pits away from the foundation in the upstream side of the river. Let the water keep collecting in these pits. Keep pumping the water away. Concrete the foundation in dry condition by keep on pumping the water at least up to two hours after finishing the concrete. Care must be taken to see that the water does not make its own channels to other side of dam through wet concrete. Dewatering pits are made on upstream side of dam. If these pits are made in downstream, excess water would pass through wet concrete while dewatering and concreting are in progress and may make permanent channels in concrete.

Some portion of the dam foundation is still left where the water level is much higher than the foundation level. This also includes the river portion to be tackled a bit later. In the above process, define the proposed diversion channel and in this portion concrete work up to designed level is completed

Create a water channel deeper than the existing river bed level with bunds on both sides for the river water to flow in this diversion channel at a length of about one kilometer on both sides of the dam with water proof lining of some material like HDPE at the bottom and the sides of this channel. Joints of the HDPE lining should be welded properly in a water tight manner. Some stone pieces are kept at the bottom and sides of film are embedded in the Earth at the top of the bund. The film is lighter than water and if not anchored properly in ground, is likely to get washed away when water is released in diversion channel. Now divert the river water flow through this channel. River bed level at start of diversion channel has to be sufficiently higher than level of concrete in dam.

In most rivers, the top layer is pervious soil and water passes through this soil, easily raising the ground water level. Hence, it is a possibility now since the river is flowing in a guided and impervious channel, groundwater in the remaining area to build the dam foundation goes down and concreting is possible by above methods. In case it is not possible to control the water flow in the foundation, for that much portion only, the well point dewatering system has to be adopted.

However, diversion channel has to be made and provision of waterproof lining could be decided at site.

9.7.2 Well Point Dewatering System

In this system, the first excavation is done up to a level just above the groundwater. A series of small pipes, called wells, about four meters in depth with non-return valves each at the bottom, are fixed around the foundation area. They are connected through header pipes and then to a heavy suction pump. When the suction pump starts, it pulls out the water from each well and throws it away. Due to this process, the ground water level in this enclosed area and surrounding areas go down. Simultaneously, the excavation progresses in the foundation area. As the water level goes below the foundation level, concrete work is carried out keeping foundation dry.

If the depth of water above the foundation level is more than two meters, multiple series of well point pumps are fixed after every additional three meters' depth, or as is felt necessary by the project engineer to draw down the water up to a level lower to foundation level. This system is patented and generally available on rent. The dam is raised uniformly in the entire length, except the diversion channel, to keep the river water flowing.

Once the dam is completed up to a top level, diversion channel is concreted as under.

Form work of a minimum of one-meter height with the top reasonably above the water level is fixed on the upstream sides of the dam in diversion channel area. Both sides form work will have plenty full of 5mm holes for the water to pass. Concrete is started in upstream side by direct dumping from one corner in a thick layer of preferably half-a-meter thickness and three meters' width, and completed up to other corner in upstream. Put another layer of half meter thick and two meters wide on this concrete by this water should stop going through this channel. Complete the layer in entire portion of the dam. By this time the formwork on upstream would be opened and fixed for next lift. Start next layer by normal concreting process. Care must be taken to ensure speed of concrete has to be more than rising of water in dam. Else, close the diversion channel in two parts which is not desirable and considered bad planning but if unavoidable, technically still ok.

Alternatively, first make about two to three meters wide wall like structure in upstream of the dam touching to dam wall by dropping M20 grade dry mix filled in woven nylon bags and Tremie concrete to create a wall in place of formwork on upstream of dam.

This work of closing diversion channel is done when water flow is minimum in the year. Before closing the diversion channel of the dam the entire submergence area and site should be cleared for submergence with written consent from appropriate authorities in writing and ensured no damage to life and property is done by rising water in catchment area. Everything would just parish within a few days in submergence/ catchment area and water would be permanently filled in the catchment/ submergence area of dam. In this process, the dam is completed.

9.8 Airports

Building a new airport is more of a logistical problem. The selection of a site depends upon a few factors such as the master development plan of the town, state or country where one should look ahead at least for forty or fifty years. Or a specific purpose - as some times, airports are built for defense use, or for heavy goods to be taken to a specific landlocked location for industrial purposes where it is cheaper and faster to build an airport instead of roads. Some of these industries include gold, oil, gas and other important mineral reserves would prefer an airport to road. The Komo Airfield, was built in Papua New Guinea, South Pacific, over the years from 2010 to 2013, in very difficult logistical conditions, and technical challenges. It was a green field project for short term use of transport equipment for gas fields, faster and far cheaper to new road in that terrain.

Building a new airport requires some issues to be resolved, such as the selection of the location of the airport, land acquisition related problems, soil conditions, if they are suitable for the building of an airport and the topography of the area.

9.8.1 Selection of Location

Technically, it is possible to build an airport anywhere. But for a large project, the soil conditions cannot be ignored. At least the soil should be suitable for cut and fill as per specifications without any special treatment like sand or chemical dozing. However, a few runways have been built on piles foundations. All these are exceptional and expensive cases. A few international airports are built by dredging and reclaiming sand from sea. Airport built by reclamation of coastal area could be cheap and fast option. No land accusation problems and reclamation by dredging is cheapest and fastest method. Gravity fill, the fill by sand mixed by water and allowed to settle with its own weight in saturated condition, does not need any compaction. Aircrafts can land from sea-sides without any fear of obstruction by high rise buildings.

9.8.2 Topographical Errors

Many large projects have topographic surveys done by aerial surveys, or are copied from Google or similar facilities which provide indicative information. Adjustments in final levels are made to make the cut and fill equal to each other. Layout plans are prepared accordingly and sent to the project site for execution. If terracing plans are based on such information, sometimes, the actual ground levels do not match the drawings. Ground level by aerial topography could be top of tree. It is done by sending waves of different types to the ground. They touch the object which obstructs the path and then this goes back to the echo sounder. The time needed for travel of these rays is noted and converted in distance. In such cases, which are unavoidable, while executing the project, some areas should be slightly lifted or levels will be reduced to match cut and fill quantities. The Komo Airfield project had over 30 meter cut and a fill of 30 meters. While execution is in progress, the level of the runway was lowered to compensate the shortage of the fill material. Google also provides indicative levels with accuracy of a few meters if ground is baron.

9.8.3 Construction Planning

An airport cannot be commissioned without the necessary services being installed. Infrastructure building, procurement, installation and calibration of airport equipment and services are main and important job. While making a time schedule for the project, the delivery dates of services should be fixed first. Accordingly, ordering of equipment is planned and implemented Suppose the project construction stretches over thirty-six months, each of the following activities would be planned as per the following sample table:

Event	By the end of
Statutory approvals, mobilization, soil testing and basic construction material at site	3rd month
Construction drawings at site	1st month
Ordering of cables, pipes, conduits, etc.	4th month
Mobilization of earthwork equipment	3rd month
Start earthwork in terminal building, power house and fire water pumping station	4th month
Placement of orders of all airport equipment	6th month
Complete Earth works up to foundation level for critical buildings	9th month
Start construction of essential buildings	8th month
Delivery at site of cables, pipes, etc. (first lot)	8th month
Start laying underground services, like cables, fire water pipes and drainage	8th month
Finish construction of essential buildings	20th month
Delivery of the airport equipment at site first stage	16th month
Delivery of the airport equipment second stage	20th month
Start erection of equipment in buildings	21st month
Complete Earth work for runway up to sub-grade including drainage	20th month
Complete Earth works up to sub-grade in taxiway and parking base	22nd month
Complete Earth works up to sub-grade in service road	24th month
Complete earthwork in car parking, etc.	26th month
Complete fencing	25th month
Complete underground services	24th month
Start erection of equipment external services	27th month
Complete runway including wearing course	30th month
Complete taxiway and parking bay	32nd month
Complete erection of equipment	33 months
Testing and commissioning	35th month
Checking and certification	36th month

Earthwork

On the project site at start of project, the first priority is to start earthwork and resources are mobilized accordingly, first being basic camp for a few officers and staff and second field laboratory for civil works. Earthwork should be done in stages and the work front should be released for the start of concrete foundations, drainage, fencing and trenches for laying underground cables and other services like firefighting, drinking water, sewage disposal etc. Laying cables of different types is the second biggest time consuming activity after earthwork. Fencing and cable trenching cannot start without site grading. Once cables are spread, heavy equipment is not allowed in their vicinity to avoid damages, which are difficult to trace, if not noticed at the time of occurrence of damage and its location is reported. These are a few problems to be addressed while planning schedule of earthwork for an airport.

It is not necessary to complete earthwork and release entire area for fencing, trenching and foundations but areas should be released for further work in agreed stages after completing leveling to a reasonable level of finished earthwork. Cables of different types and services run very close to runway and taxiway including crossings in concrete ducts to be done at or around sub-grade level of runway and covered by sub-base before further work of base course, asphalt work etc. are taken up on the runway.

9.8.4 Navigational AIDS

Navigation of a moving object is providing direction for movement and assistance to the driver as required during travel for reaching the destination safe and comfortably. An airport is equipped with such equipment which provides assistance to the pilot of an incoming aircraft start from identifying the airport from a long distance to safe landing on runway and then guide to the parking bay of the airport. These services provided by airport are called navigational aids. Installation of this equipment on an airport to the extent is decided on the basis of traffic density, visibility conditions, logistic conditions and importance of airport. There are airports which do not provide any assistance for landing on their airstrips, not even a watchman on regular basis he comes at the time of an incoming aircraft and ensures air strip is free of any stones, dogs, cattle etc. Pilot in the aircraft comes with support of his GPS- ground positioning system, makes a few rounds to satisfy that no animals on air strip and then lands. Passengers arrange their own transport from air strip to their destinations.

There are also airports, which can take over aircraft landing system from pilot and navigate from ground for safe landing of aircraft with zero visibility and adverse weather conditions. All the airports have facilities in between these two extreme cases. System and equipment installed at airport for providing such facilities are called **Navigational Aids**.

Let us deliberate Navigation Aids for a fairly equipped airport. Navigation Aids are controlled from the control tower of the airport at a high place from where runway and its approaches are clearly visible. Navigational aids need uninterrupted power supply from a power station where incoming power supply is connected to airport in addition to a few generators to automatically take over the supply to airport in case of failure of power supply. Just to understand magnitude of work a few photographs are provided of equipment on an airport. The DVOR (DOPPLER VERY HIGH FREQUENCY OMNIDIRECTIONAL RANGE) is the device to provide first indication to an Incoming Pilot, that there is a runway, in the vicinity, along with its distance from his present location. This works through Counterpoise Antennae for transmission and receipt of signals

DVOR Counterpoise and Shelter

The localizer is the heart of the Instrument landing system. It actually guides the pilot with a non-visual aid, around the centre line of the runway. The localizer antenna consists of a set of 14 Antennae, which takes multiple references at various angles to ascertain the centre of the runway. There are times, due to low visibility; the runway center line is not visible to the pilots.

Localizer

Guide slope

The Role of Glide slope in Navaids is very crucial. It determines the angle of descent of the aircraft. A little error on this might push the plane away from the touchdown zone.

Guide Slope

Guide slope guides the aircraft to descend at a proper angle and correct if required for landing on touchdown area

Fire Station

The Fire Station is the Nerve Center of any airport. It has to be strategically located, so as to adhere the ICAO standards, "International civil aviation organization", an agency of United Nations which codifies standards for airports construction and safety, 3 Minutes cycle [One Minute to reach the location of plane crash fire, the second minute to set the Laser jet aiming system and the third minute to discharge the foam]. In the event of a plane crash the operating controller or the air traffic controller raises a crash alarm, which gives simultaneous alert to the Airport Terminal Manager, Airport Security, Fire Station and the hospital. And fire is tacked in three minutes of plane crash.

Aler Room

Navaids Server

9.8.5 *Aeronautical Lighting*

Aeronautical lights are the lights in different colors, connected in parallel for uniform illumination, provided on airport to help aircraft in landing and takeoff, identify the airport, runway for landing and guide its movements to parking bay after landing. First they give green signal for airport being ready for landing. They start almost one kilometer away from airport and guide the aircraft to land by HIALS lights precisely in the centre of the runway. PAPI lights does the precision trimming of the aircraft by giving red and while signals on both sides of runway. All the PAPI lights have to glow white at the time of aircraft landing. Touchdown area is separately identified. As the aircraft is still running to stop, with different color pilot is advised when coming to danger zone and he has to stop. Further he is guided properly in different colors at strategic locations for departure.

- View of CCR for Airfield Lighting

Other Engineering Facts of Airfield Lighting:

The PAPI lights are very important to land the flight safe and correctly on the runway. It is used to help the pilot to get a correct angle and centre point on the runway by illuminating the lights RED and WHITE as standard. The HIAL lights are used to help the pilot to identify the runway to land the flight safely. Two numbers of circuits control the HIAL. The SFAL lights are used to get the centre line of the approach area for landing the flight and will be flashed for few seconds interval. So that flight can be landed safely. CCR is used to control the airfield lighting system. There are different ratings of CCRS. 20 KVA CCR is used for runway edge light circuit as it has more lights. 10 KVA is user for HIAL and Taxiway Light circuits here. 5 KVA is used for both PAIP light circuits. MAG Sign lights are controlled by only one circuit in this project

Airfield Lighting

Runway Edge Light

Taxiway Lights

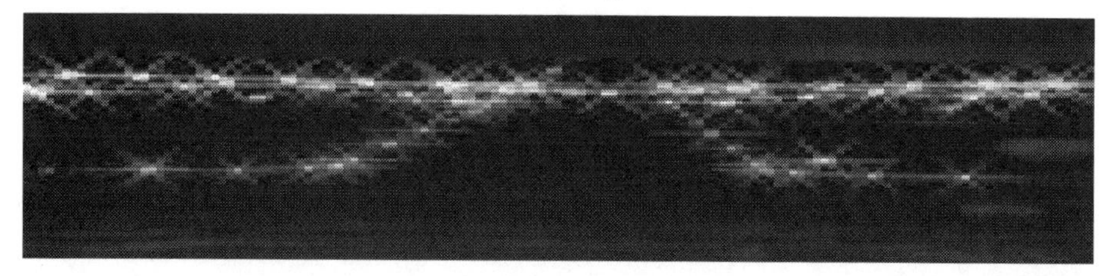

Turning Node Lights

PAPI Lights

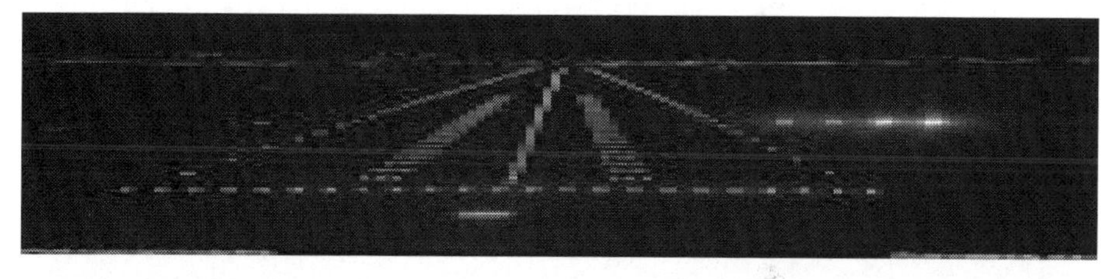

ANTONOV 124 Landing on Airport

All above facilities are part of facilities needed for an airport. This equipment is made to order and further many changes are made at site, all time consuming activates, hence have to be ordered carefully and expeditiously. Pre-delivery inspection and vendor's presence at the time of installation and commissioning are essential.

9.9 Sea Ports

Selection of Sea Port

The first thing to be done is to build a seaport close to nearest road link for export of goods or connect the industrial area to the new port with a new road. Mostly this investment is found cheaper to long distance haulage by road as per existing poor and insufficient facilities. Further bigger is the ship, cheaper is cost of sea transport provided enough cargo is available at both the ends. For bigger ships, bigger ports are required; hence there is always a demand for up-gradation of existing port facilities or building a new port. For planning a new green field port, the following are the considerations:

- First establish the type of cargo, its availability in terms of quantity per month and the duration of availability.
- The nearest coastal area for building a port and availability of infrastructure, including a road and its condition are ascertained.
- Size of port is decided for handling the cargo in terms of the tonnage per month.
- Prefeasibility study for the port location and rough cost estimates are made. Comparative costs estimate with existing facilities considering cost of port including interest on investment and contingencies are paid off in twenty-five years.
- Approvals from authorities for building a port along with arrangements for investment are taken.
- If approved start detailed prefeasibility study and then feasibility studies are carried out for construction of port.

9.9.1 Prefeasibility Study

Before going into detailed feasibility study which is expensive and time consuming a prefeasibility study is carried out as under. This prefeasibility study defines the boundaries of requirements and creates a bench mark to detailed feasibility study.

- A short drive along the coast and check if it is a straight coast or some projections of shore line towards sea i.e. at some portions the coast line is in form of C. in such cases projection of land in sea along the coast provides a basin effect and port could be built inside such basin, with reduced waves and turbulences.
- Tidal differences in sea for the area are ascertained.
- How are the waves and their normal and maximum height?
- Sea bed profile is flat or steep slope?
- Sea bed is rocky, muddy or sandy and approximately how deep is rock or no rock at all.
- Take the soundings, sea bed levels with a chain and boat.
- If the coast line is in shape of C or U, a port along one of the arms of C, or a corner could be possible if there is enough water available within about 50 meters from shore at that location.
- Sea bed level beyond this location is preferred to be sloping towards sea.
- Depth of water in low tide at the location would be ideal draft of the ship to be parked, after dredging about one and half meters at the parking bay.
- In case the coast line is straight, a finger jetty with break waters may be the solution but cost indications for breakwaters and related infrastructure could be provided only after detailed model studies and feasibility report.

9.9.2 Feasibility Study

Feasibility studies are the same as model studies explained earlier are carried out for a project in engineering laboratories with facilities for model studies. Generally, they have the topography of most regions of the world. In case more information is required, an expert team from the laboratory visits the site and conducts studies and incorporates the data in their model. Model studies provide details of orientation and location of port, turning circle for ships, entrance channel for ships, recommendations on handling siltation, dredging and need for breakwaters if required. Now, computerized studies are also possible.

9.9.3 Ports Construction

9.9.4 Minor Ports (Jetties)

Minor ports are built for fishing trawlers, passenger boats and other boats/ ships having shallow drafts say five to eight meters. They are built according to local logistic conditions. The method of construction starts with the entire area up to deep water in sea being surveyed first and the location of the jetty and its depth are decided. If the jetty on the mouth of the river connecting to the sea, first checking is to be done for siltation also called sandbar, just at the meeting point of the river and the sea. It is difficult to dredge the sandbar since it gets filled fast after dredging. Therefore, the water level at the sand bar or any water channel away from or around the sandbar is a deciding factor for the size of ships entering the river and the depth of jetty. Based on the survey data and the requirements from the jetty, the design is prepared and approved. In these ports, berthing platform is built parallel to the flow of water along the shore line, where enough water is available and connected to the shore either by a bridge or filling the entire area behind the berth with stabilized slopes of backfill.

Minor ports could be built by either of methods.

9.9.5 A Continuous Wall in Front of Jetty

A small jetty can be built by placing precast concrete blocks with interlocking arrangement if enough water is available on banks of river or shore line. Method of construction could be as follows

- The width of foundation of the berth built with precast blocks at bottom of the wall is half the height of wall, (safe) and wall thickness at top 50 to 75 centimeters.
- Layout the foundation on sea/river bed is done keeping one meter extra with on both sides to stabilize the foundation.
- Remove all the loose material from this foundation area.

- Make the area level by filling aggregates in deeper pockets. It would be better if leveling could be done by laying jute bags filled with M25 dry mix concrete.
- Place precast interlocking concrete blocks on this level foundation and build up to top on jetty front and both the sides of jetty up to shore line.
- Try to place precast blocks in low tide and put a thick 1:3 cement sand mortar in all the joints at least 10 mm thickness.
- After all the blocks are placed fill up the vertical holes as shown in drawing with concrete and a thick rebar say 25 mm dia.
- Make a coping beam on top. Necessary bollards at a few places are embedded in coping beam and other support infrastructure is provided.
- Fill up on back side of the wall entirely and your port is ready for completing necessary statutory formalities and commissioning.

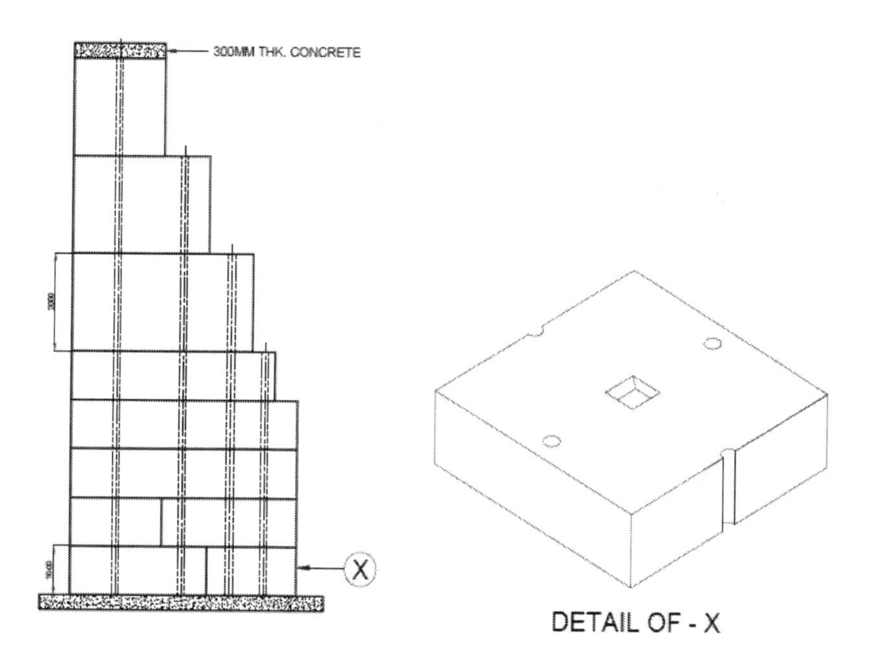

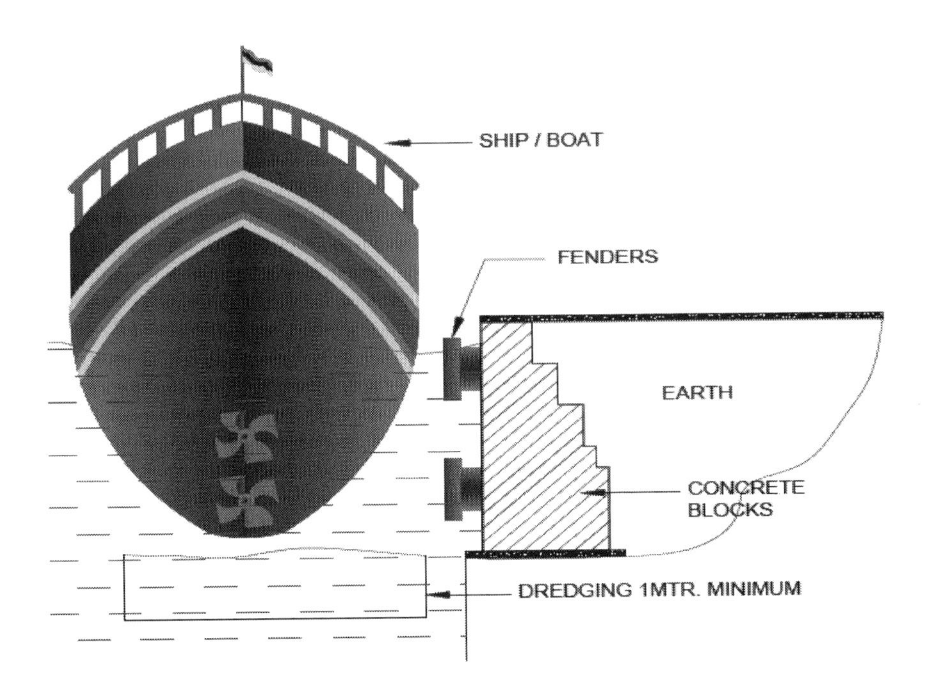

9.9.6 Sheet Pile and/or Touching Concrete Piles Wall on Water Front

Sheet piles of adequate size, thickness and bending properties are commonly used to create a sea wall/berthing facilities for ships. This wall will have water on one side and Earth fill on the other side. The ship comes and berths against this wall. The area behind the wall is filled with good Earth for loading and unloading of the ship. Sheet piles as explained in the section on bridges have a long slot with a neck on one side and a bulb on the other. While driving piles, the bulb portion of a pile, slides in the slot of other pile which provides a good interlocking of the joint between both piles and this interlocking is continued for all the pile joints from one end to other of the wall. This gives a continuous, well-knitted wall.

Earth fill and loads of cranes, goods and such else try to push the wall towards the water hence held back by a group of tie rods as shown in the drawing. On back side of tie rods, they are hooked to another row of piles of shallow depth and fitted tightly with bolts and nuts. To build this jetty, a bund is created at the location. It is filled with good Earth and is free of boulders. Piles are driven through this bund that works in a dry condition and after piles are driven and tie rods are installed, Earth on sea word side is dredged out. There would be a continuous pile cap on both rows of piles and the area is concreted to serve as deck for ease of operations.

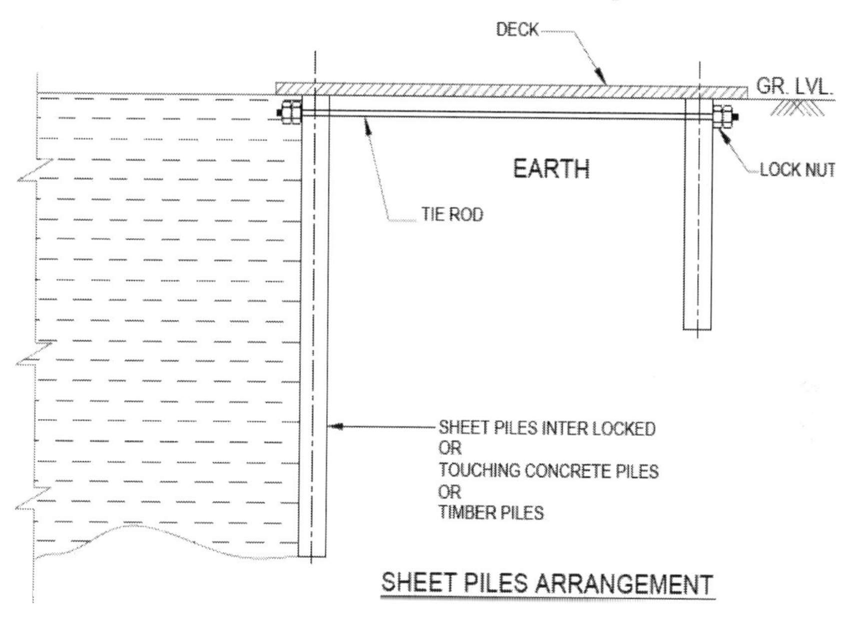

Touching Timber Piles

This is an age old method of port construction and the improvised version is use of sheet piles instead of timber piles

Like sheet piles, these piles also need back anchors to avoid bending. Since they are not interlocked, thick horizontal runners on both sides are provided at minimum three places with horizontal and cross bracings. Only seasoned wooden logs that do not decay with drying and wetting alternatively, are used.

Cast *in situ* Diaphragm Wall

Casts *in situ* diaphragm walls are similar to touching bored cast *in situ* piles with the following differences. Piles are circular, while diaphragm walls are rectangular, with a key to join the adjacent sections. Instead of a bailer and a chisel, this system has a rectangular bucket, called Kelly, with a width equal to the thickness of the wall to be built and is connected to a guide of diaphragm building rig. It moves up and down with a fixed shaft attached to the bucket. Therefore, the surface of the section of the diaphragm wall is neatly finished. It is preferred to use a Bentonite solution for the construction of the diaphragm wall for a better finish. The diaphragm wall can have vertical beam supports at designed locations by making additional concrete walls on the back perpendicular to the main wall in short distances say two meters each for stability and strength of the concrete structure. This additional supports are called counterforts; one would study in design of retaining walls. By installing counterforts, a row of piling on back side for anchorage of front wall is not required.

This method of construction is simple and lasting for long hence preferred if possible. This needs regular consolidated ground for construction, while sheet piles could be driven through make shift bund or even straight in deep water depending upon size of crane in use.

Diaphragm walls are also used for building big basements for commercial and residential buildings in restricted areas instead of retaining walls in RCC construction.

9.9.7 In Cases of Deep Water is Far Away from Shore

In the cases discussed above, back side of the port is filled with Earth. This Earth platform becomes the working area for the operation of the port. In certain cases, enough deep water is not available within a reasonable distance from the shore line and it is either not allowed or is not practical to fill up with Earth in long distance. In such cases, the port berth of a required size and depth is built at a feasible location, and is connected to the shore with a bridge. The Approach Bridge and the berth at the end of the Approach Bridge are built on cast *in situ* bored piles or small caissons/ wells foundations at equal intervals, all connected with precast pre-stressed beams. We have elaborated caissons or wells, foundations and precast concrete beams bridges in bridges sections of this book. Only the layout and purpose are different but construction methods are the same. In case of pile foundations, there are two ways. Make a cluster of piles with a pile cap at fairly long intervals with a jack up rig. It is fairly long and much easier to handle precast beams are placed between pile caps and then deck on top. This method is cheaper and faster, but needs a jack up piling rig. On the Approach Bridge, there are two or three piles in a cluster driven by a jack up rig in diameters that are more than a meter each. These piles in the cluster are connected by a common pile cap with or without cantilevers on both sides. These pile caps are placed about twenty to twenty-five meters or even more apart and precast beams are placed on these pile caps. Beams are covered by precast slabs or cast *in situ* slabs with a concrete screed on top for easy movement. This is very important to keep precast slabs together.

In a similar way in matching configuration berth and mooring dolphins are constructed. It may please be noted that lesser the number of piles cheaper is the cost of project. Instead of precast beams, structural steel built-up beams with heavy anti-corrosive epoxy paint could also be a good solution. But steel corrodes fast, if it is frequently dried and wet in water. Proper painting is of prime importance.

A port constructed by piling with jackup rigs for building foundation piles with long spans on both directions at Hazira, India.

9.9.8 Port Construction With Piling Gantry

In this case, the jetty front and approach are built on piles, at about three to five meters apart. These piles are built *in situ* by a piling gantry. The gantry standing on two rows of piles moves forward on the piles top as they are built by a cantilever structure of gantry for the next row. There are piling derricks for each pile in the row with independent piling equipment mounted on the gantry. All piles are bored and concreted before moving to the next location. There should be separate crews for each derrick and effort should be made to finish all piles in the row together. After concreting of all piles in a row, the gantry moves forward at a distance equal to the spacing between the piles to make the next row of piles. Soon after all piles are completed in a row they are connected to piles on back by strong straight and diagonal bracings. These bracings should be fixed before piling gantry moving on these piles to provide them adequate support against bending due to moving heavy loads of the gantry. The gantry moves on the rails fixed on piles and wheels are fixed at distances equal to the spacing between the piles. The gantry wheels would always be on piles and clamped after positioning the gantry.

Above each row of piles there are two wheels in the gantry one each standing on pile immediately behind pile under construction each and another pile behind. On these piles rails are fixed and the gantry moves on these rails with its wheels, if there are three rows of piles there will be six wheels and for four rows there would be eight wheels in the gantry. Gantry moves by pulling by chain pulley blocks or winches.

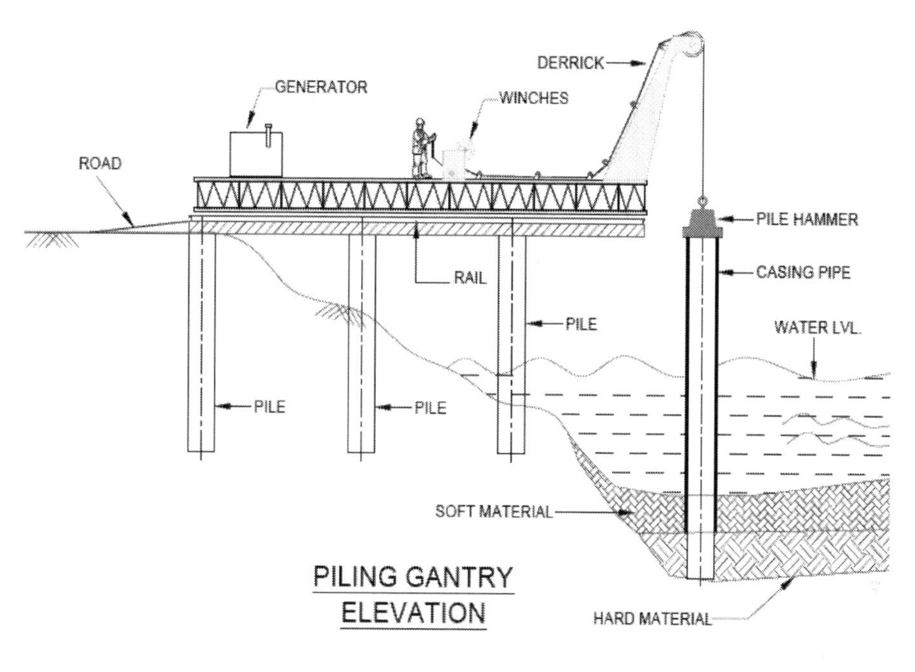

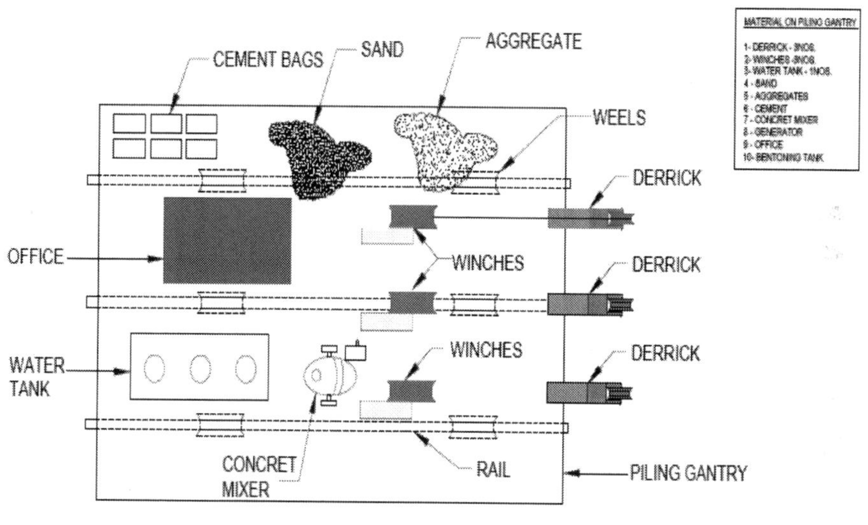

Construction procedure

- Once the configuration of the piles is decided, the layout plan for piles is sent to the site. In the meantime, the engineer looks around and identifies available gantries and their piling configurations. Efforts are made to modify one of the old gantries. It saves time and money for manufacturing a new gantry. If minor changes are required in configuration of piles to use an old gantry design, design engineer is consulted. Only on his approval, the layout is changed. Otherwise, a new gantry is fabricated, while design work is completed and construction drawings are issued. Simultaneously, other resources are mobilized.
- Once the piling gantry and drawings are ready, *in situ* erection of the gantry begins.
- Along the centerline of the Approach Bridge, the gantry location is marked on the ground.
- Rails are fixed at the top of the piles locations and levels on the ground, firmly. The layout of the first row of the pile is done on the ground or the water, in case the rig cannot drive the first pile directly from the firm ground. A few temporary piles are planned at the same spacing or at half the spacing. The piling gantry is assembled on the rails. All the piling equipment, including water and Bentonite tanks, concrete mixers etc. are fixed on the rig, keeping minimum load on the cantilever portion of the rig. Unlike the jack up rig here, piling is done with bailers and chisel operated by a winch through derricks instead of a regular pile driving machine due to light weight of these equipment compared to regular piling rig.
- Once ready to start, the piling rig is moved into position to the location of the first temporary or permanent pile as the case may be.
- Check that the **gantry is in the correct position on the new piles** by dropping a rope from the pulleys of all derricks. Ensure that the **gantry is in correct alignment. If not, correct** it by adjusting the wheels and rails.
- Lock all the wheels in their locations.
- Lower the casing pipes in the location of the piles and hammers with the bailer evenly on all sides slowly, and drive vertically in the location of the piles up to hard strata.

- Check that the casing pipes are vertical otherwise give small blows as required making them straight. Start pile boring with bailer and remove all loose Earth inside the casing pipe and beyond. When the bailer does not get loose Earth, take the chisel and deliver continuous blows for a few minutes. Then, remove all the broken soil pieces.
- Repeat the bailer and chisel operations till the hole is bored to the required level or strata.
- In case the pile is to be anchored in the rock, the rock is to be broken by chisel up to three times of dia of pile, and then concreted. This is called socketing of the pile.
- If the sides of the pile bore are collapsing, fill up the bore with Bentonite solution in water with a density that is more than the density of the undisturbed soil.
- Once boring is completed, lower the reinforcement cage in the bore after checking the depth of the bore and the length of the cage. The cage should be a meter above the top of pile, when lowered.
- Cut the casing pipe with a gas torch at the designed top level of pile. Drop the Tremie pipe in the bore and concrete up to top of pile. In the process, bad concrete is wasted by spilling over the pipe while concreting.
- Weld steel bracings in between the piles all around to make them stable, about 500 mm below the top level of the pile. Discard all the bad and good concrete above the top of the pile.
- Bend the re-bars gently outwards, not with a hammer but by bending the tool after concrete, if it is fairly strong and does not peel off on the sides of the pile while bending. Fix a set of wheels with their base plates on the new piles. By this time, 24 hours after concreting should be over. If not, wait for a few hours.
- Extend the rails of gantry up to the new piles. Move the gantry slowly with chain pulley blocks or similar tools to the next position, and continue the operation for the next row of piles. In the meantime, remove the now free rails and wheels due to the movement of the gantry, clean and service them for use on next location.

As the gantry moves forward:

- First, the pile caps are concreted for individual piles
- Place precast or cast *in situ* beams
- Deck construction follows
- Generally spacing of piles is kept in the berth, and the approach is the same. But, the width of the berth is double or triple the size of the Approach Bridge
- Piling rigs move forward on the berth
- Wheels are turned by 90 degrees with the help of jacks and the placements of equipment including derricks are adjusted
- Now, the rig works at 90 degrees of its earlier direction and completes half the piles on one side of the deck
- Again, the wheels are turned twice. It completes the remaining piles of one side of the deck
- Piling should be continued for the entire port without dismantling the rig as shown in the sketch. In the meantime, the berth is approachable and may be a part of the deck is also completed
- The superstructure is completed according to the drawings and the port is ready for certification and use. Sample movement of gantry is shown below

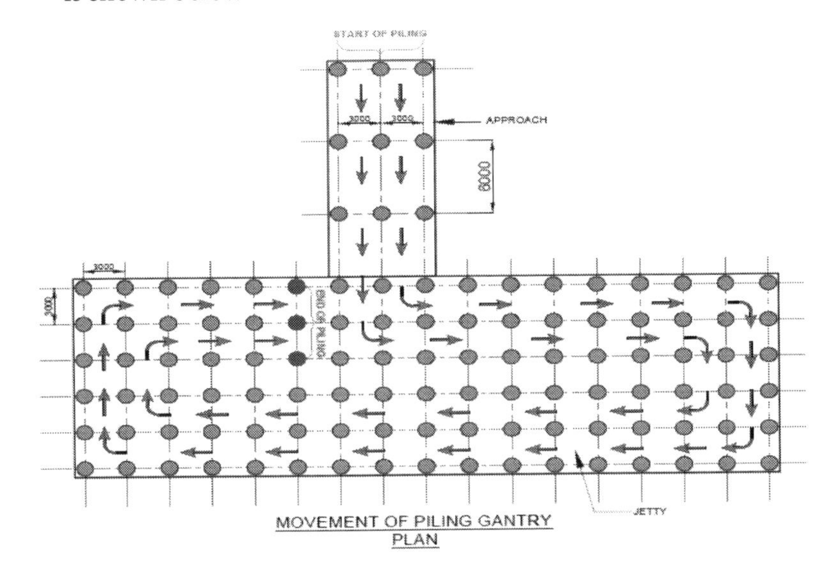

MOVEMENT OF PILING GANTRY
PLAN

This method may take time, but needs the least resources.

Oil jetty with Approach Bridge on pile foundations built in Vadinar refinery by a piling gantry for **ESSAR Group.**

Prominently it may be observed that this jetty is in a natural lagoon seen in low tide and very intelligently placed in its location. Exit channel for ship is in the direction of ship parked on the jetty. Safer for ship movements in high tide only, i.e. Maximum 10 hours waiting.

9.10 Major Sea Ports

Construction of ports in protected environment like basins and rivers is limited to shallow depths, so smaller ships can be served by these ports.

Now, with heavy volumes of cargo movement and competition between shipping companies, bigger ships in their capacity are designed and built. Commercially, the movement of cargo with bigger ships is much cheaper. Now, ships carry more than 200,000 tons of cargo and need more than twenty-four meters of deep water for sailing against 10,000 or 30,000 tons carrying capacity of ships needing 6 to 8 meters of water, are recent developments in much less than a century. Accordingly, infrastructure is required for berthing these ships on ports and to create facilities for quick loading and unloading of cargo and their warehousing. These ships need still water free of turbulence and waves along the berth for safe parking otherwise they will keep on hitting the birth hitting and damaging both the ship and berth. Further if ship is shaking it is not safe for movement of workers from shore to ship and handling cargo.

Therefore, a big area of sea in deep waters is covered and made like a basin by constructing walls in sea all around keeping enough openings for movements of ships, these walls are called breakwaters. These breakwaters provide an obstruction to movement of high waves to avoid their effect on inner side of the break water. The intension is that with help of these breakwaters, area of sea inside these breakwaters become like a lake with no turbulence in water. However, water level keeps on rising and going down as per tidal conditions. High water level of sea is called high tide and low water is called low tide. Water goes up and down twice in a day.

Proper designing and layouts of these breakwaters is important for them to be effective in breaking waves, and avoid siltation inside the basin this designing is done by a few engineering consultants around the world backed by extensive model studies and there should not be any compromises in time and costs for this important engineering design activity. Ariel view of Visakhapatnam Port with its breakwaters is shown in first chapter of this book.

9.10.1 Construction of Breakwaters

Breakwaters are of Two Types:

- Breakwaters connected to the mainland built by the end on the method for construction. In these breakwaters, one end of break water is connected to shore and can be constructed without floating crafts. However, as breakwater reaches deep waters, floating crafts can also be used to increase the speed.

- Island breakwaters that are not connected to the land and the entire construction is done by floating crafts.

For both breakwaters, rocks are required in substantially high quantities in different sizes. For one of the breakwaters in India, a rock was carted from a quarry more than 200 kilometers, by truck. It is important to find a suitable quarry as close as possible for heavy stones that do not have high water absorption and abrasion properties. Before tendering for a breakwater, all related challenges can be quantified and added to costs.

For breakwater construction, rocks are generally required in different sizes such as:

Quarry run, free of clay	Grade E material
Stones up to 200 mm	Grade D stones
200 to one ton each	Grade C stones
One ton to three tons	Grade B stones
Above three tons up to ten tons with defined percentage of ten tons and above stones	Grade A stones

Maximum size of the stones is determined on the basis of the current and wave height and these stones are placed on top layer and get continuous beating by waves and high currents. They should not be disturbed by waves. A few centimeters of marine clay in the breakwater alignment is okay, but if the marine clay is more than fifty centimeters, it is preferred to dredge the clay before building the breakwater. But, a little settlement of breakwater is not of much concern. So, many designers do not force the dredging, but adequate compensation should be given to contractor in quantity measurements of fill material sinking in the marine clay.

It is always advised and hardly implemented in initial operations – that it is not advisable to break big stones with secondary blasting. They are required towards the end of the job. Many times, towards end of project quarrying is continued in large quantities for the sake of a handful of heavy stones at the cost of the contractor. If the geological strata are such that heavy stones are not available, concrete blocks in the form of cubes for low waves and tetrapod's in the case of higher waves are used instead of big stones. This helps break waves and do not slip or get washed away.

Construction of Breakwaters With the End-on Method is as Follows:

- Lay out the breakwater on land with two or three marker poles erected in the centre line
- Probe the sea bed to find some marine clay
- Take the sea bed levels
- Bring a crane on barge with a grab to remove the silt and throw it on both sides away from the bed of breakwater called side casting
- Bring another crane on the shore with a long boom and a dragline bucket
- Start dredging and side cast marine clay in about ten meters long patches for the full width of breakwater by barge crane
- As the dredging proceeds, start dumping E grade material (free of fines of clay and silt) with dragline, before the area is filled with clay again. If enough depth of water is available, E grade and grade D material could be brought in the bottom opening barge and dumped in section
- This is followed by dumping on the shore and bulldozing of higher grades of stones of C grade above grade E &D material up to a level above water level in order crane could walk on these stones
- As the work proceeds, construction goes to deeper depths and works of placing grade E and grade D are placed by bottom opening barges
- Bulldozers continue placing Grade C stones up to a little above water level
- Crane on land moves back and start arranging Grade B first and then Grade A, or the tetrapod's as per the drawing on sides and top of breakwater

- Thus finish the task from the land end and move towards the end of the breakwater till finish
- Fines are removed by a big screen called grisly as shown in following drawings. Truck load of grade **E** material mixed with fines is brought on grisly and material is dumped on the screen. all the unwanted fines are screened and good material is picked up from front of grisly

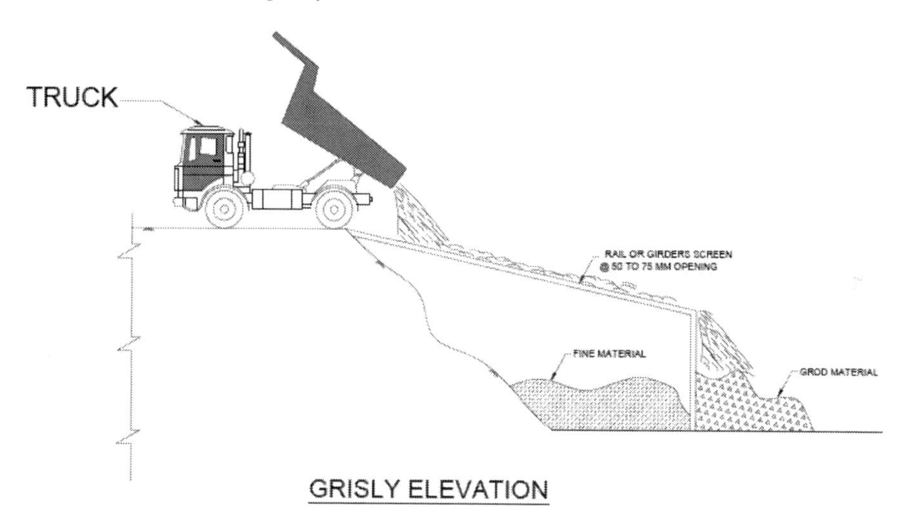

GRISLY ELEVATION

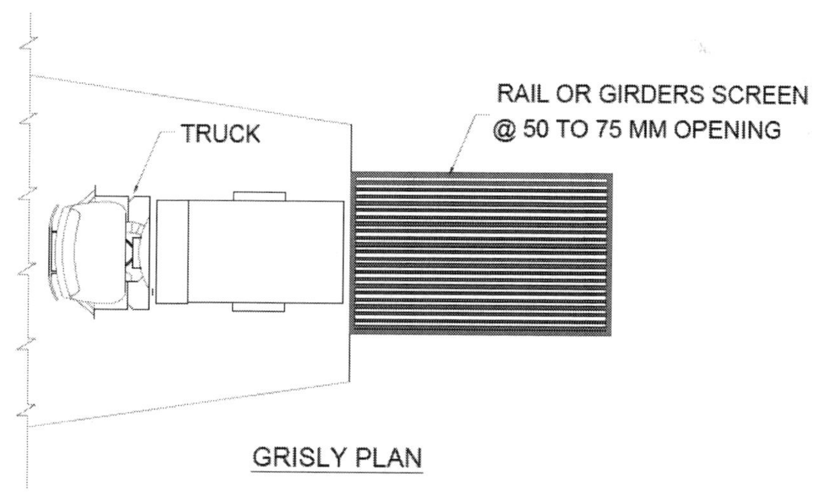

GRISLY PLAN

Island breakwaters

Island breakwaters are constructed in the sea as islands and not connected to the shore. The construction procedure is similar to the end on method, except that the entire construction operation is carried out by floating crafts. Each craft is fitted with self-positioning systems. The location of the craft is always known and monitored. The bottom opening barges are self-propelled barges about 200 to 1500-ton capacity. They are of two types: barges with bottom opening gates with chain and winch operations, and split barges, which are fabricated in two halves and joined together by two hinges, and is operated by hydraulic jacks. In order to fill the barge, it is closed. Two halves touch each other at the bottom and look like ordinary hopper barges. To open, press the lever of the hydraulic jacks, the bottoms open up wide and all the material falls down instantly.

For parking and filling these barges, a few temporary jetties are built at loading points. Grade C stones seen in water at low water level, grade B and tetra pods are being placed by crane walking on wooden pads on an island breakwater in Kakinada, India shown in the picture, longest island breakwater in Asia at the time of construction built by ESSAR Group.

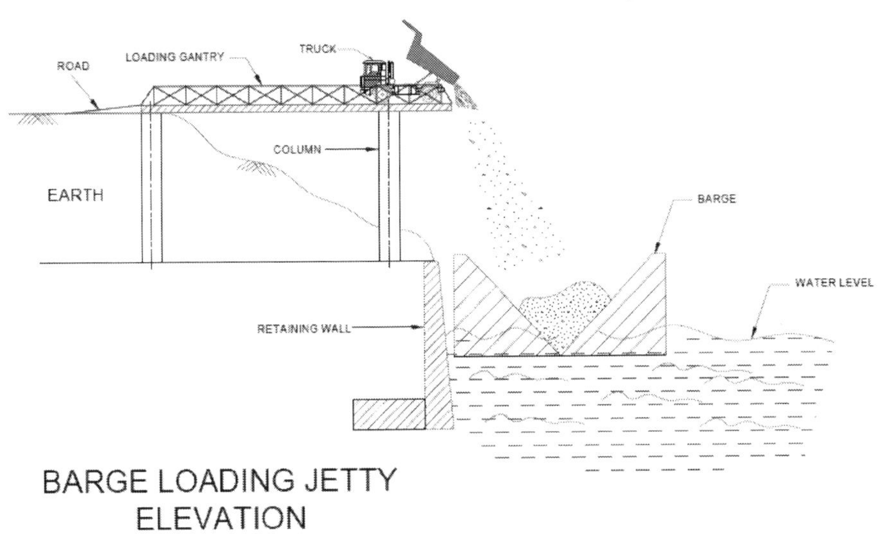

Drawing of barge loading jetty, barge is being loaded with E grade material

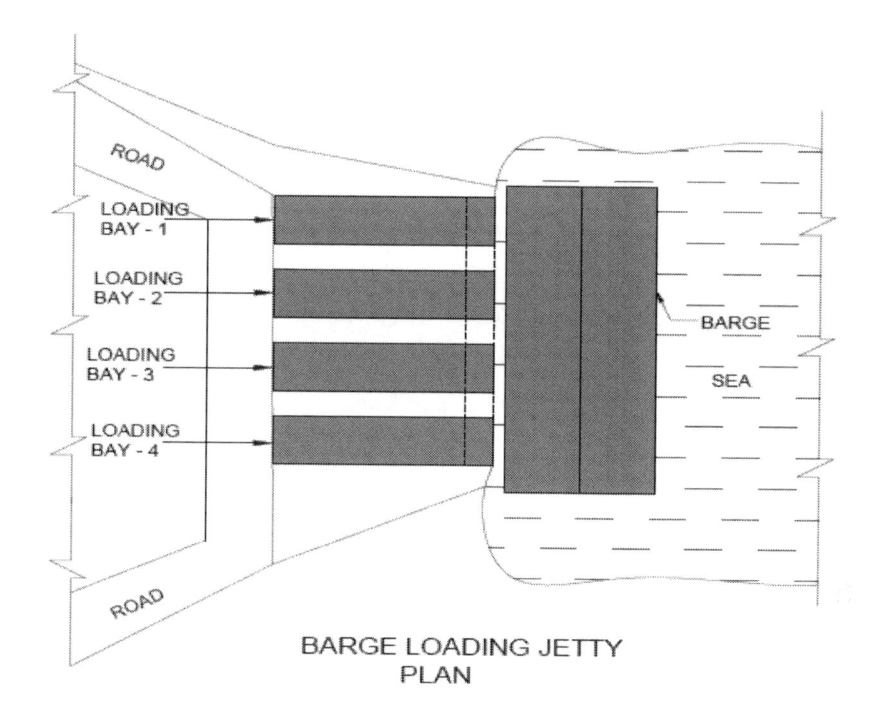

BARGE LOADING JETTY
PLAN

For loading higher grades, barge hoper is provided with thick lining of timber to avoid damages by falling stones.

9.10.2 Construction of Deep Water Port/Harbors

A cluster of deep water and shallow water ports called harbors are constructed in manmade basins covered around by breakwaters. Extent of these basins is usually a few kilometers length and breath. According to favorable logistic conditions many ports, called berths are built in one harbor. Necessarily some amount of dredging is required for safe movement of ships inside the harbor. This dredging is called capital dredging. Dredging for maintenance of depth by removal of silt coming in harbor is called maintenance dredging.

Deep water ports having greater water depths could be built by either of the following methods or their combinations.

1. Cast *in situ* bored piles connected with precast or cast *in situ* elements
2. Sheet pile wall on the water front
3. Touching cast *in situ* piles for the length of birth
4. Cast *in situ* diaphragm wall of RCC as water front with RCC superstructure construction
5. Precast concrete caissons connected with pre-stressed concrete or steel bridges
6. Precast concrete caissons touching each other
7. Steel pipe piles connected with steel beams and cross beams covered by reinforced concrete deck slabs

While the construction techniques and procedures are similar to bridges and minor ports, except quantities of items of works, more sophisticated equipment, skills and risks of all kinds are much higher, hence to be built by careful designs and execution with good expertise and safety considerations.

9.10.3 Precast Concrete Caissons Connected With Pre-stressed Concrete or Steel Bridges

This is a frequently used option for construction of ports that are either parallel to the shoreline with the ship berthing on one side, or a finger jetty with ships berthing on both sides. Caissons are placed at about twenty to thirty-five meters' clearance between them. All the ship berthing equipment including fenders and fender walls are fixed on top of caissons.

These caissons are interconnected with concrete or steel bridge for the continuation of services like water and fuel pipes, and the movement of personnel and goods. The photograph below is of Marmugao port iron ore berth built by ESSAR where concrete pre-stressed beams are placed in a bridge between caissons for working platform.

On each caisson there are two concrete columns on back side to support the steel bridge for movement of iron ore ship loader. In photograph, a ship is being loaded iron ore, standing alongside the port.

9.10.4 Precast Concrete Caissons Touching Each Other

In cases where the required depth of water is a bit far from the shoreline and the reclamation of the land behind the port is permitted, a thick wall is built along the face of the port. All the area between this wall and the shore line is reclaimed. Two ends of this wall are also connected to the shore line with lighter walls. This closes a basin from all sides. This basin is filled with locally available soil starting from one end and well compacted. Care has to be taken that a few gaps are kept in the basin for exit of water. The front walls to be used for berthing the ship will have deep water as per requirement water for berthing of big ships is made out of precast concrete caissons of depth as required touching each other while placing in their foundation. Slip form concrete for these caissons is preferred since number of caissons is large enough. The photo graph shown below is one of the ports in Mumbai, India built with this system. Here, the pre-casting of caissons up to five meters was done by slipforming system in a temporary dry dock and balance building up the caisson by slipform concrete in floating condition. After placing all the caissons in position, at an appropriate time gaps between them are closed by concrete.

9.10.5 Steel Pipe Piles Connected With Steel Beams and Cross Beams are Covered by Reinforced Concrete Deck Slab

This is the fastest and most durable mode of construction for a port, provided that all matching resources, including the material, equipment and men are available. It requires good fabrication and a painting shop of about 2000 tons per month of fabrication close to a seaport and should be committed for the project. A fleet of barges with tugboats are required for the transport of fabricated and epoxy painted items on the site. A few heavy barge cranes fitted with powerful vibro-hammers should finish the piling for the project in a hundred days. A few barge cranes for the erection of steel structures of the port, starting from the thirtieth day of the start of piling and finish the erection within twenty to thirty days after the piling ends. Meticulous planning is done for all activities, including the supply of fabricated and painted items at site as per schedule.

After the detailed layout is finished, a few barge cranes with vibro-hammers move into the site along with the first load of the pile pipes. Start driving the piles to finish all piles in a hundred days, in a sequence. A good vibro hammer with skilled operators should drive twenty to twenty-five piles in twenty-four hours. Simultaneously, a gang of gas cutters are used to cut off the extra pile of pipes above the cut off level. Fill the piles with concrete or sand as per design requirements, concrete is preferred, in case due to any reason pipe pile starts corrosion, concrete in the pile would help in carrying load. While driving through harsh strata, epoxy paint on pile could erode or break. This would not even be noticed by engineers and pile would lose all its strength within short period. Hence as a safety RCC concrete is recommended inside the steel pipe piles in addition to epoxy paint.

Fix prefabricated pile caps on these piles. Start the erection of the steel structure on the pile caps from one side, followed by precast slabs. All steel structures should be welded and fixed rigidly with each other. Concrete the portion ready with precast slabs placed in position. This area, after concreting, is the base for further activities to be done conveniently. Fix the fender wall and fenders, services and deck equipment.

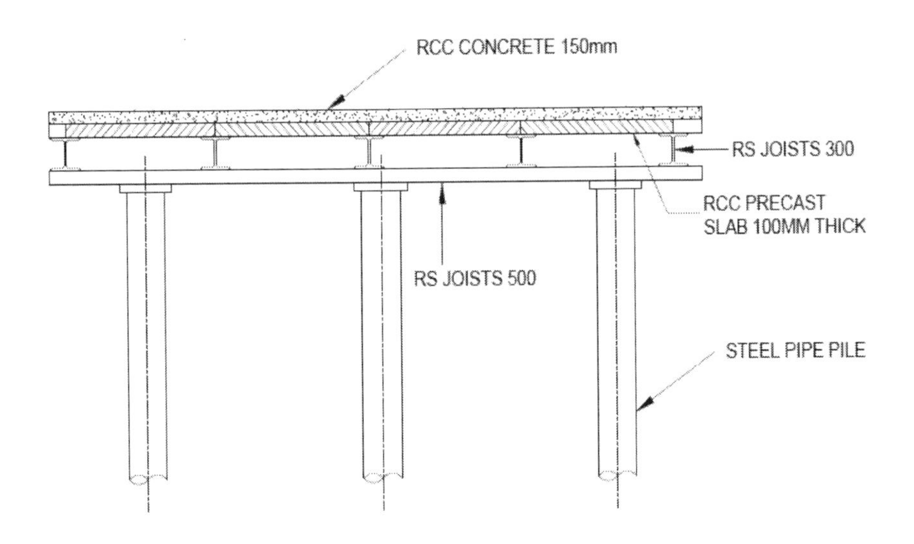

All the wildings on epoxy coated steel structures and areas heated by welding should be treated properly by a painting scheme recommended by paint manufacturer.

9.11 Industrial Projects

Building a new industry or extending an existing industry is always very interesting. It is a different type of challenge for the construction Manager. Here, the contractor works for an individual, an entrepreneur or his representative, depending upon the size of the organization and the project. He has to provide technology, quality and aesthetics in the best possible manner that is available in each of the fields, in the best price, the least time and best quality.

Interesting to note that value of construction contract is only 20 to 25% of total cost of the industrial project.

The balance amount takes care of land, design, process knowhow royalties, cost of equipment, finances and overheads of the owner. This 25% component of cost is main driving force for successful completion and timely commissioning of the plant. In case of abnormal delays, the owner could impose penalty on contractor maximum up to 10% of contract value which not affordable for the contractor. But this 10 % of construction contract is only 2.5% of total cost of project which is nothing compared to losses to the owner and in abnormal delays it would be difficult to absorb by the owner due to loss of production, increased interest on capital investment and loss of tax holidays committed by government on timely completion and commissioning of the plant and many more. Reputation and reliability of contractor are always main factors in decision making of award the project to a contractor and selection of his Project Manager.

Thus, the processes of planning a project and costing are very important. Mostly, projects are funded by a financial institution and they keep an eye on progress and costs while releasing funds. Cost overruns beyond a reasonable sum becomes difficult to be paid to the contractor by the financial institution, even if owner is willing to pay.

9.11.1 Project Planning

For an industrial project first broad scope of work is estimated depending upon past experiences and unit rates for each activity are worked out with the contractor based on overall quantum and nature of work. At this stage only process is shortlisted and approximate area of land at the particular location are only known factors. Even soil investigation is to be carried out at choice of contractor, time permitting before deciding unit rates of different types of foundations.

Once contractor is decided, he starts fencing of the area and start site grading, cutting and filling soil including compaction to required density and discarding unsuitable soil in green belt and disposal areas.

Simultaneously, the manufacturing process of plant is finalized, equipment is ordered and general arrangement of all the components of the project are frozen and compiled in the form of a document known as project specifications and plotted on a set of drawings known as general arrangement drawings. These drawings and specifications become the basis of project planning. By these time soil investigations are also completed and types of foundations for buildings and equipment are also frozen. However, in almost all the cases these drawings are revised many times but by now there are some documents to work on.

On the basis of these general arrangement drawings first any changes in site grading as required are made and instructions passed on to the site engineers. With the general arrangement drawings and project specifications following additional information is available to project planning engineers

- Approximate quantity of concrete to be poured
- Type of foundations and their quantum
- Quantity of steel structures and their types to be used
- Quantity in volume, number of equipment with volume, height and weight of heaviest equipment to be erected
- Approximate scope of mechanical, electrical and instrumentation services
- Indicative schedule of release of construction drawings and delivery of equipment at site
- Any other information necessary for project planning

On the basis of above detailed planning of resources, time and costs are worked out. Still in some ways, construction of industrial projects is easier compared to those discussed earlier since,

- All activities are generally land-based
- Inventory and procurement of very small items is done in logistically convenient ways, since all activities are at one place

Industrial projects have their own problems and work gets smoother if they are addressed properly and timely.

- Land acquisition
- Finalization of the process
- Construction drawings for the project are released in an agreed sequence
- Finances are tied up for the project

There are hardly any projects where any of the above activities are completed before the contractor puts his foot on the job site, since the owner does not have patience and does not waste any time idling and tries to do whatever is possible, while other things are sorted out. In such circumstances, it is important that project execution and planning be a joint effort of the owner, the consultant and the contractor. Invariably, everybody tries to protect reason for the delays on his part by blaming others and ultimately the contractor has to take the blame for all the delays.

Hence, the contractor should not have a very aggressive start, since if he fails to complete the first few activities in the agreed schedule, others will have an excuse for not doing their part. This has a snowball effect. The project gets delayed. The resources remain idle, increasing costs to the project and adding to the burden of unnecessary correspondence.

If the contractor is ahead of his schedule, it would build pressure on everyone and ultimately everybody would do his work faster to avoid getting blamed, which will result in the work being completed sooner. Approach roads are the most important resource for efficient working. Therefore, it is preferred to build permanent roads inside the plant up to a sub-base level, as far as possible, and then these roads would be used as main haul roads for the construction of the project and get fully stabilized by the time the project is complete. There are always instances where services cross under the road. Locations of crossings should be planned. Small culverts should be built while building the roads for crossing of services under the roads.

For foundations, if bored piles are prescribed, the agency responsible for piling should also be responsible for clearing material that has been removed from the pile bore simultaneously, as piling proceeds and disposed at designated spoils dumping ground to avoiding slush spreading everywhere. Deepest pits, foundations and trenches should be taken up for construction before the surrounding area is addressed. Efforts should be made to have the working area approachable until work in any particular section is finished. Detailed planning should be jointly discussed and all activities including the owner's scope included in the schedule and monitored by contractor with threats of extra clams if delays are due to consultant and the owner.

9.11.2 Site Grading

Site grading is one of the first activities to start on a project in the following sequence;

- First identify the area reserved for green belts
- Prepare temporary approaches if required from the plant area to the green belt
- Remove the top soil in the whole working area and spread evenly in the green belt area.
- Start digging from the higher side of the plant area and filling in the lowest areas. As one section is completed, take up other areas as per the schedule.
- For lower areas, adequate drainage facility should be installed before moving to the location for piling and such else.

9.11.3 Piling

Design of pile foundations are based on soil investigation reports and are very important for the stability of all structures. It is always recommended that one pile should be built outside the foundation area and tested for failure. Results of the pile test are then sent to the designer and he can make changes in the number of piles in the individual foundations. This pile is called **TEST PILE.** Piling for foundations is to be carried out according to the sequence that is provided in the schedule. When the test pile is being cured and testing is being done, piling work on project foundations should continue. The design engineer keeps a provision of extra piles in a few foundations. These extra piles are to be driven in case of a failure of the test pile. A foundation should not be covered with a pile cap before the test pile is tested and test results are approved.

9.11.4 Deep Pits for Underground Tanks Required in the Project

Sometimes, deep underground and waterproof concrete tanks are required for some process of the plant, like a deep tank to hold hot Zinc in liquid form for continuous hot dip galvanizing of steel plates. These pits are much deeper and have to be water tight difficult to build in normal routine work procedures, hence should be addressed as follows:

Check to see whether the ground water table is above or below the bottom of the pit that is to be constructed by making a bore. If the level of the ground water is below the bottom of the pit, work can be done without a problem and the job can be done easily with proper side protection. Side protection is provided by providing timber planks as support with horizontal struts (supports) between both faces of excavation on all sides, step-by-step as excavation progresses. They are removed as the structure starts coming up and is back filled in stages. If ground water level is higher than the bottom of the pit, piling in surrounding area should not be done until concreting in the pit is completed and cured. Since water level is higher than the bottom of foundation, extensive dewatering would have to be done, pulling water from the surrounding area to enable pouring concrete in a dry condition. This would also disturb soil configuration in that area and soil providing friction to pile may become loose by movement of soil particles, hence reduction in bearing capacity of pile. Piling should be done after concrete in the pit structure is completed.

9.11.5 Procedure for Dewatering for Foundation

- A number of holes are driven, about one third of length or width whichever is higher extra below the bottom of the base slab and three to five meters away from the outer side boundaries of the pit for dewatering on all four sides of the pit.
- Put a temporary PVC pipe casing in the bore holes to protect soil falling on pumps in operation and then lower the pump in the PVC casing.
- Start excavation about two meters extra on each side as place required for fixing formwork and other activities. Strong shoring to be done as excavation proceeds.
- Before excavation reaches ground water level.
- Start the pumps and throw the water quite far in an attempt to lower the ground water level in the area. If water is thrown just a few meters away, it would go down fast and start circulating makes dewatering difficult.
- Keep on excavating as water level goes down, until the bottom of raft is reached and continue the pumping day and night continuously.

- Concrete the raft when dry and put water bars at joints including where walls would start.
- Continue the wall concreting with proper backfill on behind and removing the shoring planks in stages as concrete goes up. After concreting of walls is done fairly up to a level above ground water level, stop dewatering slowly, and one by one pump should be stopped with two hours break for each pump.
- At the time of stopping dewatering, unit weight per square meter of concrete mass on the foundation of the pit should be much more than unit weight of soil at that level in surrounding area. Else the surrounding soil would push foundation up to strike a balance of uniform loading in the area. This is called upheaval.
- While stopping the pumps, if a leakage is observed, restart the pumps and pressure grout the leaking joint and wait for two days. Pressure grouting is done by chipping holes at points from where the water comes from. Fix small 20 mm nozzles with cement mortar. Inject cement and water mix grout till refusal with a high pressure grout pump. Once the pit is completed, checked for leakages, fill it up with water up to top to reduce at least thirty to forty percent of external ground pressure on slab and walls.
- Start piling in the surrounding area for other foundations ten days after all pumps are removed, with water remained filled in the tank to reduce Earth pressure on its walls by piling in adjacent area.

9.11.6 Concrete Foundations and Trenches

Concrete foundations are generally on piles but trenches for cables and pipes are always on compacted soil. It would be advisable to make trenches along foundations which are at lower levels than the foundation first. Then, the foundation work adjacent to the trench should be taken up. Similarly, trenches above foundations at the bottom should be built after the foundation is concreted and the surrounding area is backfilled and compacted. Further care has to be taken to maintain approaches for all works till the particular work is finished. These trenches could be damaged by movement of construction equipment.

In case of a need to allow traffic over the foundations, first, wrap the exposed foundation bolts with thick cloth soaked in grease. Second, cover the whole foundation top with the M10 colcrete up 50mm above top of bolts. Make sure not to bend the bolts. After this, the sides should be filled by Earth properly and compacted. After this, spread gravel on the prepared bed, and the road is ready. For cranes and excavators, long and strong wooden pads are necessary over this temporary road but in any case excavators and cranes must not stand on foundations and work. In all cases, care has to be taken that foundations are not overloaded, moved or tilted due to this unplanned loading not considered in design of foundation.

9.11.7 Steel Structures Fabrication and Erection

9.11.7.1 Fabrication

For all industrial projects, a major part of the superstructure is made of structural steel. Depending on the logistics and technical reasons, either the steel structures are fabricated in a commercial fabrication shop and transported to site in road transportable sizes or are fabricated at a makeshift fabrication shop at or near the site. For fabrication at site, the following steps are to be observed.

- Plan out the fabrication shop and write down the method statement for fabrication and transport of fabricated structural members to location of erection for approval of the engineer.
- Get approval of the fabrication shop proposal by the competent authority as directed by engineer. Start procurement of shop equipment and their installation.
- Prepare shop drawings for structural work that is to be fabricated in the shop on the basis of structural drawings provided by design engineer. Work out the quantities of material required along with preferred sizes to minimize wastage. Order the material confirming to material specifications in lots and deliveries as per requirement and fabrication schedule.
- Arrange checking and certification of welders, fabricators, riggers, supervisors, cranes and lifting tackle. Purchase welding electrodes compatible to chemical composition of fabrication material and their sizes.

- When the first lot of material should arrive in the shop and stack them properly and segregated according to sizes on wooden sleepers. Check test certificates and other quality checks. Start cutting the steel with least wastage and carry out edge preparation for welding as per specifications. Testing equipment and personnel should be posted at the shop by this time.
- Fabricate the structures as per shop drawings. Make a trial assembly of a few sections to check the dimensions, angles and matching holes. After fabrication, checking and certification send fabricated sections for shot blasting first and then painting.
- Shot blasting is a mechanical process to remove even small traces of rust before painting. Small metallic nodules mixed with compressed air are bombarded on the steel surface and by hitting with good force continuously and all the rust is removed. Steel surface turns from brown to grey, the original color of steel. Shot blasted member has to be wiped clean and immediately a coat of primer of paint is applied before exposure to atmosphere. This operation has to be done in a fairly large closed chamber with proper clothes and breathing gear for the workmen.
- Sand blasting used earlier but now banned in some countries. In this process sand is used instead of metallic nodules. Sand breaks in pieces which may pass through breathing mask and hence a health hazard.
- After shot blasting and primer painting, full painting is to be done; identification number to be marked and ready for dispatch. Painting technology has been developed to a quite advance stage different paint manufacturers are marketing their products which are claimed to be more effective and cheap in different environments in which the structure has to exist. The environments could be sea water, harmful chemical storage or fumes or any other location needing protection of surface. They suggest three or more coats of different paints in specified thickness and sequences.

9.11.7.2 Loading and Transport to Erection Site

- For loading and transport of painted steel structures special care has to be taken so that the painted steel structures do not come in contact with any metal, with which if structure is rubbed, paint may be damaged. Secondly, while painting, structures are kept on thick timber planks without nails and rapped in gunny bags. Members of the structure are to be lifted by certified cotton or nylon belts and placed on truck or trailer again on battens by a certified crane having hooks at designated points as directed. After loading is completed, all members are tied firmly by belts on the truck. Finally, in case of haulage on roads, pilot vehicles in front and behind the trucks or trailers should escort the convoy to avoid accidents with other vehicles on the road.

9.11.7.3 Erection of Steel Structures

- Before erection, a safety briefing to be given to all those involved by the engineer and the safety officer. The safety officer should check and confirm the erection site and see if it is free of obstructions and is properly barricaded.
- Erection engineer should check to see that all items to be erected on the date are within the safe reach of the crane. Check the loading charts and the safe capacity of the crane to work at a particular height, and that the radius and position of the crane on firm and safe ground is available for crane to stand and lift structural for erection. It is important to check if there are enough bolts, nuts, washers and other fasteners are ready on site with their matching spanners.
- The safety officer to check lifting tackles, scaffoldings, safety belts and shackles are enough in good working condition for the maximum load to be lifted.
- If erection is at height, too much sun and wind should be avoided and the timing of erection should be modified accordingly.

- Once everything is ready, the crew is fresh and not tired, start after the erection is signed off and erection work is done as per requirements and stopped by the evening.
- Erection work at night should be avoided as far as possible. However, if unavoidable special care is a necessity for a good crew and enough lights are arranged.

9.11.8 Erection of Equipment

- In industrial projects, some equipment is erected in covered sheds as well as in open area as per their utility and operational requirements. Equipment in sheds are erected after shed and foundations for equipment are completed. For erection of equipment in shed, an overhead traveling crane is provided of an adequate capacity to handle and erect the biggest size of equipment in a single lift and further maintenance of such equipment.
- If the overhead crane is not available on time, purloins and roof covering in a few spans are delayed to facilitate the erection by other crane available at site.
- For erection by a normal crane, in open space an erection schedule has to be prepared to facilitate erection of all the equipment in safe reach of the crane. It should not be first come first erected basis. Otherwise for equipment far away from crane reach will be difficult to erect.
- Sometimes equipment erected have to be dismantled for movement of crane and erection of heavy equipment. This has to be avoided by erection in correct sequence.

The following checks and detailing are necessary for erection of heavy equipment on their foundations for smooth operation without any problems.

- Check that the foundation location and levels are accurately and differences are noted.
- Difference in levels can be adjusted by chipping the top of foundation and filled up by epoxy grout after equipment is erected, and approved for grouting. If the location is shifted by a few

centimeters, equipment is erected at the location as it is, only if it does not foul with other equipment and connections of services are modified accordingly. In case of major shifts, layout of all the foundations in the cluster is checked accurately and modifications as considered appropriate are carried out before start of equipment erection in the region.

- Top surface of foundation is chipped and aggregates exposed. Clean the top with air water jet.
- Make the layout marks on foundation clearly visible. Check that foundation bolts are correct and threads are not damaged. Grease the bolts after cleaning with diesel and keep nuts handy.
- Keep adequate shim plates handy for level adjustment minutely.
- Hook up the equipment at identified lifting points by the crane.
- Bring the equipment on the foundation and lower slowly. Care must be taken that crane does not swing with load over any personnel and still swing has to be very slow.
- After erection of equipment, it is leveled, put required shim plates, tighten the bolts and release the crane. Grout the foundation with epoxy grout or as specified after approval.

9.11.9 Installation of Services

Installation of services are carried out simultaneously as construction work is progressed but taking care of damages while other activities are carried out together. Power, communication and instrumentation cables with sufficient tag numbers for identification at a later stage at the time of joining or termination are laid in cable trenches. Fuel, water, and other pipes as required including firefighting system are laid in the specified trenches including intermittent pressure tests and other quality checks. sewage lines are laid and connected to sewage treatment plant. Important to note that rest rooms including proper connection to sewage treatment plant, commissioning of firewater system and electrification are the items to be completed and commissioned before the commissioning team steps in the plant.

On one of the projects erection team refused to enter the building since rest rooms were not properly functional till alternative arrangements are made.

9.11.10 Commissioning of Plant

Commissioning of the plant is carried out in stages as the work progresses and it is important that different units of the project gets completed and are made ready for commissioning as per agreed schedule which is based on process and stages of commissioning.

9.12 Residential Buildings

In residential buildings, the main role is of the architect, who plans all the facilities. The engineer has to give a strong structure and install facilities as planned by the architect. The main difference between residential building structures and other projects involving high skills is that in buildings, good looking finishes are more important. Therefore, the engineer has to build a strong building, especially high rise buildings which can withstand beatings of rain, wind and most important earthquakes. Taller the building, more are risks and more care has to be taken for building the structure starting right from foundations to roof slabs and other components like walls, doors, windows etc. In a high rise prestigious residential building in Mumbai, all the outer edge of balconies is not in a vertical straight line. Such things are well noticed by common man looking from a distance.

Still for looking good and conveniences, buildings are extensively plastered with good finish. Plaster on concrete should be by providing steel wire mesh reinforcement after thorough cleaning the surface of dust and loose particles. A thin layer of plastering compound on the concrete surface would give more bonding with concrete. The engineer must understand once concreting is done, a thin layer of mortar is seen on all over the surfaces. Cement does not stick permanently without additional aids. One alternative is to scale out this mortar layer by chipping with a hard tool and expose aggregates everywhere area is to be plastered. It would be a matter of pride and achievement if a building is built in good quality. In India, the state headquarters of Chandigarh was built without plastering. The Lotus Temple in Delhi is another example of good workmanship in recent structures.

Antiskid tiles laid in proper slopes in rest rooms, speak for themselves. Similarly, all the sewage, drainage, water and firefighting lines should be honestly hydro tested and fixed properly to avoid leakages. Tiles in all the rooms are to be first planned on a paper to scale and then joints are decided to provide a good pattern and avoid wastage. In countries like Australia even footpath tiles are laid properly with smooth level joints. If someone fells down on the road, he would search an uneven joint of tiles, take a photograph and ask for compensation due to tipping off at this joint. Now think of quality of workmanship needed in rest rooms which become slippery when wet and have to be safe.

Doors and door frames should be from seasoned and treated wood free of knots. Make sure all the edges inside and outside the building are straight. Stairs are another neglected area. If steps and treads are not uniform, one could even fell down. External night lights should not have glare in bed rooms and defused on walls and not on windows.

LEVEL 10

Quality Assurance and Quality Controls

Quality Assurance is required at every step of the work right from the site investigations till the end of the project and good quality performance of various components of the project is necessary not only in guaranteed period of generally one year, but even thereafter. It should not happen that some vital equipment start giving problems or project is not built as desired or could be better. We buy a car, but before making the purchase, we make it a point to ask about its performance and ultimately buy one which will remain trouble-free for many years and less maintenance cost even if it is expensive. This is true in all walks of life, including engineers and companies involved in engineering profession. Good quality work does not start or finish with the contractor building a good structure as per or better than the specifications but it starts at the level of drawing concept plans by the consultant and their approvals by the owner.

The project passes through various stages and at every stage starting from concepts, detailing, designing, cost estimation, selection of contractor and monitoring the quality standers are required to be maintained by standard tests carried out as frequently as agreed in TQM plan as per ISO guidelines for the project. Proper correcting measures as per accepted engineering practices as necessary.

LEVEL 11

Estimation and Tendering

11.1 Estimation

On the basis of details provided in the tender documents and the site visit estimates are prepared. This is explained in costing and tendering in project management section of the book. Once the estimate is approved and the amount to be put to the tender is frozen. The concerned documents are passed onto the tendering engineer.

11.2 Tendering

Tendering is a very important function in contracting. It is a challange to put the estimated price in different heads of prices for the benefit of the contractor and still be acceptable to the owner. The following are some of the common considerations for the tendering engineer while distributing the price over various heads:

11.2.1 Requirements of Cash Flow

Nowadays, generally, advance payments are discouraged. It requires a bank guarantee which is an extra cost and a hassle. Secondly, the contractor finds a way to pull out the advances from the contract before the job starts, which is a further strain on the cash flow for the Project Manager. The contractor has to spend considerable money initially before the first work bill is raised on the client. Tese costs are such as mobilization of construction equipment by new purchases or in house transfer from other projects, mobilization of manpower, camp facilities, office facilities, performance guarantees, insurance of works and construction material for at least two months. These are termed as priliminary expenses.

It is advisable that cost of all the priliminary expenses are individually worked out and charge to the client as these activities are executed. These costs are not more than 7% of contract value but one can charge easily upto 20%, if not more in preliminaries.

Some of items added are like mobelization of Site Manager at site,establishment of site office,submission of quality document, submission of safety manual, submission of project time schedule, mobelization of equipment (equipment name and capacity has to be specified), submission of bank guarenties, submission of insurance policies. Client is also happy that he is paying after some thing is happening. Contractor gets a suport to his cash flow. Project manager is happy for he has not to ask financial suport from head office, except guarenties to bank and insurance company, which itself is a big suport, since contractor is taking all the risks and pledging his assetts to bank and insurance company.

Only safe guard is to be taken that by charging big amounts in preliminaries,enough money has to be left at disposal of Project Manager that he complites the job easily and keep on feeding head office for their overheads. Money once taken at start of project by head office is forgotten and money has to flow to head office every month.

11.2.2 Differential Pricing

Money remaing in tender after taking out for prelimanaries from the tender price is distributed to all the items of work,like excavation, concreting etc.. These prices are distributed in proportion of their costs. Again, the items of work for which quanties are likely to increase are charged more and items of works for which quantites are likely to reduce, are charged less. Further some items should be executed early are charged more. However totral of all the prices has to be equal to agreed final price decided by management with estimation team.

Positive Repercussions

Invariably, all the governments and organizations have budget allocations for various works and such allocations keep changing on all projects depending on the new priorites. Under such circumstances, towards the end if the owner has different priorites, he would intend to delay the project due to his problems. This puts contractor in a better place to bargain, since he has made his money and has nothing to lose. He can ask for reimbersement of the idling cost, the cost of demobilization and re-mobilization, when the owner would like to complete the project.

Negative Repercussions

On a project making good profit in the initial stages due to high preliminaries and front loading, if these profits are diverted to other projects, it is difficult to take it back when needed. Projects towards the end get into a starvation stage. If the contractor is not able to help by providing cash, project start slipping which has snow balling effect on losses since each delay would cost money. This gives a bad name to Project Manager and the contractor.

Escalation in Costs During Construction

As a general practice for any job prices are worked out on the basis of prevailing rates for material, labor and comodities at the time of tendering. In such cases if cost of basic components in the contractgets increased or reduced, this has direct impact on the costs and hence profitabelity in the contract.

For contracts of shorter dueration it could be understood that in such a short time period there may not be much variations in costs and any contractor could afford to absolve. In such cases he in turn enters in fixed price aggrements with supliers of major materials and other services. This could be agreed by supliers on short term basis but still if there are substantial infelations, supliers try to get out of fixed price aggrement at the cost of the contractor. This is a big risk the contractor has to evaluate and add to his contract price.

In case of long term contract running for a few years, if fixed price is demanded, contractor has to keep provisions in contract price for variations in costs during contract period and even beyond in extended time, unless it is properly documented and owner would not be willing to pay extra costs in the extended period. In such cases the price escalations are worked out by interpolations of infelation trends for the past a few years and added to the price.

As a safeguard to this anamoly government in almost all the developed and developing countries monitor price variations in major comodities including labor wages on regular basis and publish indices of prices of all the comodities on national basis and major zones, seperately. The variation in these indices are noted for a few items from such publications and escalation in costs are worked out both upward and down ward and bill value on monthly or fortnightly basis is adjusted for such variations. This benefit would be applicable to the contractor only if it is agreed and recorded in the contract before signing the cotract for the project.

In India, Ministry of statistics and program implementation monitors and publishes all India consumer price index for major commodities on regular basis and Bureau of labor statistics under United States department of labor monitors and publishes consumer price index in United States.

This operates as under

For example, contract indicates and accepted by owner, following percentages of costs for computation of escalation

Labor and staff	30% --	calculated on the basis of whole sale price index
Oil and lubricants	20%--	calculated on the basis of cost of diesel at a particular location
Materials	40%--	calculated on the basis of price index of a particular commodity

One can distribute to any number of indices as he likes from the tables published by these departments. Safeguard being that minimum 10% of tender price is considered as profit to the contractor. Escalation in costs is not payable on profits, hence total of all the percentages as to be claimed should be not more than, on 90% of bill value. Escalation in costs are worked out as under.

Esclation of cost in a catogery during the month	**A**
Comonent of total costs	**B%**
Price index indice at the time of tender	**C**
Price indice for the month of billing	**D**
Value of work done during the month	**Q**

$$A = (D - C) \times B \times Q / 100$$

Similarly esclations for all the catogries are calculaated and added to the bill for the month.

If C is more than D, esclation would be negative and bill amount would be reduced accordingly.

11.2.3 Special Conditions of Contract

There should be a meeting of the tendering engineer with all the people involved in the preparation of the tender. All possible concerns and gains are recorded and concerns are added to the tender as special conditons in an acceptable format and language. Some of these are if applicable:

- For the execution of this project, some imported construction equipment to be purchased and used. Any foreign exchange variations during the contract period and extended period will be reimbursed on the basis that 20% of the work bills would be considered foreign exchange components, converted into US dollars at the rates prevailing on the date of tender submission or thirty days prior of tender submission.
- This project is planned for time-bound completion and resources would be mobilized accordingly. In the case of delay not attributed to the contractor, he will be paid compensation for all the extra costs including loss in profit for the extended period.
- The right of way for the project would be the responsibility of the client. Any delays in the work due to the non-availability of the work front would be adequately compensated to the contractor by giving extension of time with reimbersement of additional costs.
- The contractor has not taken insurance for hostalities, nor covered in the price, hence losses due to hostalities or unrest of any kind would be assessed jointly and reimbersed to contractor.
- Contractor has assumed building materials like earthfill, sand, aggregates, stones, timber etc would be locally available in radius of thirty kilometers from the work site. If local material is not approved and contractor is required to bring material from distant places, fresh rate analysis would be jointly conducted and approved for payments before bringing such material to site.

- Electricity would be made available at site by the client on normal chargeable basis and no generators are planned at site
- The owner would provide ten acres of land of any incumbences, close to work site, well connected by road for his temporary facilities free of cost available up to one year after maintenance period of works are completed.
- Client would arrange the lease of quarries free of any charge for the contract period for such material in sufficient quantities. Government levies if any are included in the costs.
- No dewatering of ground water is envisaged in costing and would be paid extra if applicable. No rock is invisaged in excavation of foundations etc, and if need to excavate rock, it would be paid for at negotiated rates.
- A few more conditions could be added in similar lines in the tender document.

11.2.4 Tender Negotiations

Soon after the tender is submitted and opened, one gets a fair idea on the chances of winning the contract. In case of a good chance for one's invitation for negotiations, the whole document is revisited carefully for mistakes. Some managers say that you can win a contract only if you have made a mistake. On the other side of the coin, if the estimater is advised that company should get the job, the estimator starts cutting corners and the price becomes too low to work with. Before the contractor is called for tender negotiations, the team should review the tender to find out our mistakes and then form a stratigy. Here are three real examples:

When a tender was being prepared for the Nhava Port with touching caissons in Mumbai, a glaring mistake was found in the tender. A few items had mistakes in putting the decimals correctely. A quantity of 10,000 was put as 1000 or lesser. The winning contractor increased the prices for these items and proportionately reduced the final price to tender.

Ministry of education tenderin Oman, contractor got the job, since he had quoted killogram rate of re-bars for a ton of steel but accepted and finished the job with profits.

Cement sillows for cement plant in Tamilnadu lowest bidder did not had relevence experience hence likely to be rejected. Contractor offered to take payment after job was complited as a single payment and took away the contract.

One has to be very careful in tender negotiations. He is generally required to take on-the-spot-decisions. He can always ask for a few hours at least, to think about the important issues, which is not wrong, but gives a good impression to the other side. These Decisions are vital and not to be taken in haste

Once the contract has come to the stage of the award, an efficient team is selected for the execution of the project. All files of the tender and the negotiations are passed on to the designated Project Manager. The Project Manager plans for the project with the tender documents as a reference. He prepares his list of resources and cost estimates. These cost estimates are submitted to the management, deliberated in detail and then approved with some corrections. The approved cost estimates become the guideline to monitor the execution of the project under the responsibelity of Project Manager and his team.

Let's proceed, with the work with all the commitments, work will show the way to cross hurdles. GOOD LUCK!

PREM VARDHAN

Made in the USA
Middletown, DE
15 April 2017